WHAT I WISH THEY WOULD HAVE TAUGHT ME IN SUNDAY SCHOOL

About

Science, The Bible and God

Philip Singerman

Copyright@ 2007 Phil Singerman
2nd ed. 2008
Printed in the United States of America

Published by:
Grace Publications
2050 W. Devon Avenue
Chicago, Illinois 60659
773.465.5300

Book Design & Cover by:
Carlos A. Irizarry

For licensing/copyright information, for additional copies or for use in specialized settings contact:

The Brookside Groups / P. Singerman and Sons, Inc.
Phil Singerman
1-877-8NEWBOOK
1187 Wilmette Avenue
Box 208
Wilmette, IL. 60091
sales@8newbook.com

www.8newbook.com

ISBN 978-1-4276-0647-1

Several years ago, as we were walking to our car after leaving Yom Kippur High Holiday services, as is our tradition my family began discussing what we found inspiring and uninspiring about the day. I had with me my copy of the Torah (the first five books of the Bible) so that I could locate the biblical context of some of the portions that we had just read and, therefore, be able to add to the conversation. After a few minutes my middle son Joel turned to me and said, "Dad, why don't you write a book about Judaism?" From that moment, Joel never stopped challenging me until one day I realized three things: 1) I had not truly studied Judaism as thoroughly as I wished I had; 2) what I did know lacked substance; and 3) my input during pertinent conversations with my children was beginning to sound rather hollow.

In response, I devised a strategy of serious study of science, the Bible and God. I discovered volumes written about Judaism, physics and cosmology, yet I could not find an individual book that readably encapsulated Judaism's relevance, essence and structure that spoke directly to an average 21st century Jew like me. (Now there's an oxymoron for you: an "average" Jew.)

The more I studied these areas, the more intriguing and meaningful Judaism became for me. I learned that we could "connect the dots" between modern science and GOD and that the Bible was not simply a series of disconnected morality plays. I came to respect the fact that the Bible was to be read as a whole, because understanding each preceding portion of the Book is essential to recognizing the significance of the upcoming portions.

So I decided to write a book that focused upon Judaism's most important concepts and fundamental underpinnings. I wanted this book to be directed toward the mind and heart of today's typical Jew. In order to bring some order to this task, I chose the Book of Genesis as the platform upon which to base my journey.

I hope that as you read this book you experience, as I did, several "Aaha!" moments when you think to yourself, "I surely wish that I had learned this earlier in my life". As I discussed this book with many educators, family members, friends and acquaintances, both Jewish and non-Jewish, I found that the passage of time for each of us had not lessened our desire to understand what makes religion in general, and Judaism in particular, really tick and to understand why it is important, awe-inspiring and relevant.

Before we begin, a few housekeeping details:

1. There currently exists a debate among scholars regarding the origin of the Bible: was it written by GOD or by Man? For those who believe that it was written by Man, the most accepted thinking is that it was written in four different godly inspired portions. The details of this debate are beyond the scope of this book; therefore, I have chosen not to discuss this subject in this book.

2. When I reference Man, Him, He or him I am using the masculine form of the pronoun only for purposes of consistency of writing style. Please feel free to substitute Woman, It, humankind, people, She or her as you wish.

3. I cannot explain anything without using words. However, as I assign verbs such as "see," "hear" and "know" to GOD's actions, I am not implying that He has the same five senses that humans have or that these words even approximate any accurate description of GOD's awareness, communication, visage or acts. In my limited ability as a human, commonplace terms are all I have.

4. In Judaism there is only one deity. However, getting to know His complex essence takes some work. Therefore, in order to make this essence a little easier to grasp, I refer in my book to two aspects of godliness that together bring spiritual "oneness":[1]

- When I refer to GOD in all capital letters, I am acknowledging those spiritual aspects that are *beyond* Man's capacity to fully comprehend.

- When I refer to God in upper and lower case letters, I am acknowledging those moral aspects that are *relatable* to Man. I agree with author Daniel Matt, who observes that, without a personal or relatable God, there is no possibility of a true relationship with the divine.[2] It takes recognizing and studying both aspects to appreciate the extraordinary, universal godliness found in Judaism.

5. In order to clearly convey the meaning of certain concepts found in the Book of Genesis, I have jux-

---

1 Daniel Matt, *God and the Big Bang* (Woodstock, VT: Jewish Lights Publishing, 1996), 37

2 Daniel Matt, *God and the Big Bang* (Woodstock, VT: Jewish Lights Publishing, 1996), 48.

taposed discussions of comparable content even if this places these subjects slightly out of chronological order. For example, the different Covenants are discussed in succeeding chapters in this book even though they are not found in succession in the Bible. While the significance of GOD's name could have been discussed anywhere in this text, I felt that it related most closely to the "Chosen People" chapter. Further, Jacob's two most significant encounters with GOD follow each other in this book but not in Genesis.

6. This book is meant to be read without, necessarily, a Bible at hand. However, you may want to have one handy just to enrich the significance of any discussed subject. I find *Commentary on the Torah* by Richard Elliott Friedman to be particularly meaningful, and it is the source I have used for all biblical translations.

I hope that you enjoy reading this book as much as I enjoyed writing it!

Philip Singerman

Acknowledgements

Dr. Elliot Lefkovitz, Religious Director, Am Yisrael Synagogue, Northfield, IL; Adjunct Professor of Judaism, Loyola University of Chicago, IL; and Adjunct Professor of Jewish Studies, Spertus Institute of Jewish Studies, Chicago, IL. Without constant input from Dr. Lefkovitz, this book never would have seen the light of day. He graciously edited my thoughts on the theological portions of this book and patiently helped me to incorporate what I learned into one relevant edition.

Dr. June Fox, Former Dean of Graduate School of Education, Lesley University, School of Education, Boston, MA. Dr. Fox wonderfully assisted with the molding of my words into a coherent document. She was instrumental in helping me take these words and create a book that, I hope, connects the dots of basic Judaism for the reader.

Harvey Kelber, Wilmette, IL. Harvey was a counterbalance: he would challenge my assumptions, which allowed me to make sure that what I said was what I meant to say.

Joanne Medak, Wilmette, IL. Joanne kept an eye out for style and possible redundancies.

Robert Wigoda, Attorney, Chicago, IL. Robert's family is steeped in the study of Judaism. He held weekly study sessions at his law office and was kind enough to let me know his thoughts on the more "human" episodes of the book.

Avrom Fox, Owner, Rosenbloom's Books, Chicago, IL. Avi graciously kept up with my work as it proceeded from idea to finished product.

Rosanne Ullman, Editor and Educator, Better Writing Group, Wilmette, IL. Rosanne helped me to streamline my writing and polish my manuscript for the purpose of publication.

I also want to thank **Roger and Joan Gold, Dan and Sherri Ollendorf, Michael Willens, and Miriam Tuschman,** all of whom volunteered to read several drafts during this long process and to provide their honest thoughts and suggestions.

My most heartfelt thanks go to **Charlie and Mickey Silverstein.** Although Charlie comes from an editing background and Mickey is a reading specialist, it was their personal efforts regarding the creation of this book that I most appreciate. They have been with me from the initial concept to the final word and have never wavered in their support and encouragement. It truly would not have been completed without them.

And, finally, I want to thank the specific people for whom the book was originally written: **Susan, Noah, Joel, Micah, Cheryl, my mothers Frances and Adele, and Barbara, Danny, Judith, Jerry, Lynne and Jen,** all of whom had to endure unending discussions regarding its contents. They listened, smiled and allowed me to "figure this out" while probably feeling numb from the repetition. They never faltered in their support, and without their love this project would not have been possible. I love you all!

Thanks to everyone.

Philip Singerman
08/06/2007 (updated 2/17/08)

Table Of Contents

Some 20-odd years ago, as my sons approached four, six and nine years of age, I made a valiant pledge to myself: I would never use the phrase "because I said so" as an answer to questions that my sons asked me regarding religion, Judaism, ethics or morality. Rather, my responses would have to be supported by meaningful rationales that I would find acceptable and could justify and defend.

When their questioning began, their youthful challenges were relatively simple. My answers flowed easily and were forthcoming. However, as my sons became older, I found that the questions they raised and the challenges that they brought to me grew increasingly more difficult to satisfy. Still, I was holding my own.

Then one day at the age of 11, my middle son, Joel Levi, announced that he had decided to become a vegetarian for reasons of sanctity of life. On the heels of this pronouncement my eldest son, Micah Isaac, stated that he was going to follow in his brother's footsteps and, oh by the way, could I please discuss with him Judaism's stance on murder!

Not to be outdone in the challenge department, years later my youngest son, Noah Asher—yes, these are their real names—after winning the Inspiration Award in his senior year of high school football, contracted meningitis. My wife Susan and I watched over him in the hospital special care unit after we were told that he might have only six hours to live. I understood then the real challenges to faith. (Thankfully, Noah is fine today.)

The truth of the matter is that our first challenge actually began some years earlier, when Susan and I were having difficulty conceiving a child. We agreed to every test and undertook every regimen, unfortunately without success.

While on a planned trip to Israel, we stopped at the Western Wall in Jerusalem. As I approached I saw the thousands of notes written on scraps of paper that had been placed into the cracks in the wall by individuals from all over the world. I meditated for some time on those hidden words that I could not read. I turned to leave but then returned and placed a note into the wall myself as I decided that I had better look for the wisdom that would be required if we were to remain childless.

One month after we returned to Chicago, Susan and I became jubilant to learn that she was carrying our first child! This initial true challenge underscores my motivation for writing this book. Susan and I are very grateful to now have Micah, Joel and Noah.

But their questions! As the boys grew, it became clear that my religious education was no longer sufficient to provide me with the ability to wisely, or even adequately, handle the boys' inquiries regarding their life situations without going back on my pledge to not simply take the easy way out. Internally, I'd always felt that I had a "correct" sense of my religious beliefs; however, the leaps of faith that I found myself making were getting larger. I faced the choice of either covering up my lack of knowledge by giving quick answers that lacked substance, or using the reason, "Because I am the father." Even worse, with some of their questions I felt as if I did not know how to respond at all.

As I watched them grow, I suspected that even tougher questioning was still to come. The more challenging the questions, the more insecure I became regarding my understanding of the true meaning and wisdom of Judaism.

I had grown up in a kosher home and attended a typical Conservative synagogue where my parents happened to be among the founding members. I went to Sabbath services regularly, became a Bar Mitzvah and was confirmed. But even with all of this Jewish background, I clearly was not fully prepared to explain to my own children the why's implicit in the questions they were posing.

When I discussed this predicament with family members, friends and acquaintances, both Jewish and non-Jewish, I found a surprising response. They had exactly the same set of reactions that I did when talking to their families; they experienced the same feelings of inadequacy. Like me, they sensed that they knew what was "right", yet could not comfortably discuss the why's with their children. Some admitted to being dumbfounded on occasion.

Today's kids are sharp, aware and worldly, not prone to accepting doubletalk. They know when you are dancing around an issue and are quick to reject your words if they feel your answers are not thought through well enough.

All of that had me feeling somewhat uncomfortable. My responsibility as a father was being compromised as I sometimes slipped into using jargon, clichés and platitudes. I did not want to lose my boys' respect nor, frankly, my own self-respect, so I made the decision to try to find out for myself what really makes Judaism tick.

I had no idea what a task I had set for myself! The questions seemed endless.

Is Judaism or religion in general truly meaningful in today's world?

- What is the Jewish GOD?

- Can we believe in GOD when referencing modern science and current global events?

- Is the Torah relevant for us today?

- Can the Bible be read as a guide for our lives, or is it simply a collection of stories?

- Is it possible to have an idea of what our GOD might be?

- In the end, is the leap of faith necessarily so vast that making it becomes impossible, or is Judaism's GOD "close" enough to man that we can find meaningful ways to relate to His presence?

One day I was making a Shiva call, which is a tradition that sets aside several days after a death for relatives and friends to visit the home of the person who died in order to extend condolences to the family. My friend's father had passed away, and at the Shiva house I found myself in the late gentleman's study. What I saw there astounded me! An entire wall held DVDs and cassettes containing lectures on various topics prepared by professors from universities across the country. I asked my friend if I could borrow one on math. I took it home, put it into the DVD player, and after about one hour came upon a lecture on the Fibonacci number sequence.

The next few minutes changed my life!

The Fibonacci number sequence is the series of numbers that occurs when you start with the number 1, add it to 1 and then add this sum to the previous number and so on:

1+1=2, 1+2=3, 2+3=5, 3+5=8, 5+8=13, 8+13= 21, 13+21=34, 21+34=55, etc. Thus the Fibonacci sequence is 1, 1, 2, 3, 5, 8, 13, 21, 34, 55, etc., to infinity.

It so happens that these numbers display several levels of exquisite order in nature. Three particular aspects of this natural order captivated me.

First, it is uncanny how many plants, vegetables and fruits reflect the sequence:

- The buttercup flower has 5 petals; lilies have 3; irises have 3; corn marigolds have 13; asters have 21, 34 or 55; wildroses have 5; larkspurs have 5; cornflowers have 55 spiral petals; delphiniums have 8.

- Pineapples have 13 diamond shapes on the pineapple skin if you count in a clockwise direction and 21 if you count them counter-clockwise.

- Poppy heads have 13 swirls of seeds.

- Pine cones have 13 seed pods in a spiral when you count to the left and 8 when you count to the right.

- Cauliflower has 5 indentations between stalks that go to the right and 8 between stalks that go to the left.[1]

It is extraordinary that all of these numbers are found in the Fibonacci sequence.

Second, the Fibonacci sequence leads to the Golden Ratio. As you go through the sequence and divide each number by the number immediately preceding it, watch what happens to the resulting quotients:

1/1=1, 2/1=2, 3/2=1.5, 5/3=1.666, 8/5=1.625, 13/8=16.25, 21/13=1.615.

As you continue with this process, the quotient always approaches a constant of 1.618.[2]

Third, by placing blocks representing these numbers in a certain geometric pattern, you will always produce a "picture" of what is called a "Logarithmic Spiral." If you can envision the way a conch shell spirals from inside to outside, then you know exactly what this looks like.[3]

"The Fibonacci numbers are nature's numbering system," maintains adventurer Stan Grist. "They appear everywhere in nature from leaf arrangements in plants, to the pattern of the florets of a flower, the bracts of a pinecone, or the scales of a pineapple. The Fibonacci numbers are therefore applicable to the growth of every living thing, including a single cell, a grain of wheat, a hive of bees and even Mankind."[4]

---

1   Stan Grist, "The Hidden Structure and Fibonacci Mathematics," at http://www.stangrist.com/fibonacci.htm.
2   Ibid.
3   Ibid.
4   Ibid.

The extraordinary order found within the seemingly random series of numbers was spellbinding to me. Could this order exist by chance? Could the relationship between these numbers and nature be a coincidence? Was this universal elegance or just plain luck? I turned off the DVD player and knew right then that I had to try to find out for myself.

I began by devising a personal strategy. I would initially study the Torah, starting with the Book of Genesis along with commentaries of Genesis and explanatory texts. Next, I determined that if I was going to be able to talk with my boys about these subjects and their various levels of complexity, I had better have a working knowledge of physics and the science of cosmology so that when discussions arose regarding the universal significance of topics such as the infinitesimally small and unimaginably large, I would be prepared. My studies had to be comprehensive in order to determine whether the physical world and the spiritual world can coexist. Are they intertwined with each other, or are they part of a united whole? Even more interesting, can one help to explain the other?

What I found was startling!

There are many, many books with insightful, wonderful interpretations and explanations of religion, the Bible, Judaism and science. There are so many, in fact, that it was actually overwhelming. However, even with so many volumes of information in existence, I found that it was difficult for someone like me to relate to them either individually or collectively in order to create a mosaic of understanding. The information and insights that I found in these many volumes are like stardust particles: they appear in large numbers but are very much separated from each other. There's a little bit here, a little bit there, but not enough in any one place to condense and form a single star. Similarly, there is a little information here and some there, yet not enough within any single volume to which the average Jew can meaningfully comprehend and relate.

It took me several years of study to be able to write the book that is in your hands—an explanation of the basics of Judaism referencing the Book of Genesis that many people might find relevant, relatable, readable and comprehensive; a book that might be able to connect the dots for the average Jewish person or interested reader of any religion. I am also pleased to say that after studying these texts, I have come to several of my own personal interpretations that I hope you will find meritorious.

I am not a scholar in any sense of the word. I started this journey with little in-depth knowledge about these subjects and am now writing about the portions that I found exciting, wonderful and awe-inspiring. Many people have spent their entire adult lives delving into the substance of the universal questions discussed herein. These are the real scholars. I simply attempted to learn from them. The authors of the books that I have read and the professors that I listened to on DVDs became my "friends," as I imagined them debating the issues while I studied their thoughts. The exciting part was synthesizing what I learned into an understandable and relatable whole.

When I began this journey I clearly did not know where it would lead. I actually had some trepidation when I started, because I did not know how I would react if I discovered that science and religion were incompatible. What if I could not find room for GOD in today's world or if the required leap of faith grew too large to make? What would I tell my children? How would I react? Yet, it was clear that there was no turning back. I had a need to know and that was that! I would deal with whatever I found.

So I began. For my focus I used the Book of Genesis, where we find that sometimes GOD relates to Man, sometimes Man relates to God and sometimes human beings relate to other human beings. These relationships are enveloped by science, history, society and religion all helping to define the fullness of Judaism. We see that, throughout the Book of Genesis, the primary parties individually and collectively mature both religiously and personally. I, too, have matured as I have come to more fully appreciate the wonder of "oneness" and the wisdom inherent in the sanctity of my Jewishness. Additionally, I have found that humankind is truly a significant part of this oneness.

As I studied the texts and a piece of the puzzle would fall into place, I would pause for a second, relish a momentary sense of accomplishment and then think, "I wish that they had taught this to me in Sunday School." I would smile and continue, and soon it would happen again. I would come upon an interpretation, a relevancy, a connection, an explanation, an insight or a personal understanding and I would sit back and say, "I wish that I had learned this in Religious School, or surely earlier in my life."

This journey has been exciting and awe-inspiring. Most important, it has taught me that experiencing increased levels of trust allows Man to pursue higher degrees of faith. As I conclude this part of my study of the Book of Genesis, as well as the explanatory texts, basic physics and cosmology, I am pleased to say that I embrace Judaism more fully today than at any time in my life. I find it insightful, wise and relevant, and based upon actions that are supported by institutionalized hope.

Three points before we begin:

1. Through our journey together we will find that in science, as well as in religion, some degree of a leap of faith is required. At crucial junctures, the mathematics of physics breaks down or lacks concrete answers to some of its central concerns. For example, quantum mechanics is based upon probability, not absolute certainty.[5] The Heisenberg Uncertainty Principle states that it is impossible to tell both the path of a particle and the exact location of that particle at the same time; therefore, everything is, even at the level of the most infinitesimally small, a function of probability. "Any attempt to measure one will unavoidably disturb the other," Bill Bryson says in *A Short History of Nearly Everything.* "This isn't a matter of simply needing more precise instruments; it is an immutable property of the Universe."[6] Indeed, the scientific method itself is not based upon proving a theorem correct but, rather, simply proving that it is not incorrect. Therefore, a theorem can never be absolutely correct.[7]

As another example, science dictates that Dark Matter and Dark Energy exist in the universe even though

---

5   Matt, 44.
6   Bill Bryson, *A Short History of Nearly Everything* (Broadway Books, 2004), 144.
7   Stephen Hawking, *A Brief History of Time* (Bantam Books, 2005), 14.

they are not currently visible to Man. The universal laws of physics tell us that they must exist in order to corroborate the observed effects that they have upon many parts of the universe. Thus, scientists can explain much of the *how's* and *what's* of the universe without necessarily knowing the *why's*. In later chapters, we will see how this specifically pertains to Dark Matter and Dark Energy.

It is very similar for religion. Humankind fundamentally understands that we are limited beings and, as we become increasingly aware of the essence of our limitations, the thought occurs to Man that there is assuredly something larger than himself that he cannot completely comprehend. Therefore, while we do not exactly know the *why* of what it is that is larger than Man, we trust that a GOD does exist just as scientists don't know the *why* of everything they believe exists. Within this context of trust, I have become more aware of how universally all-encompassing is the Jewish religion, which has helped to shorten my leap of faith from a seemingly impossible jump to a more manageable hop.

2. There may be people in the scientific and religious fields who may not care for some of what I have written. Please remember that my work is a labor of love since, at the outset, I did not find a comprehensive book to which I could fully relate. The wonderfully inspiring writings that I found, while pertinent to my studies, did not sufficiently encapsulate within one book the subjects of science, the Bible and GOD. No single book presented the Book of Genesis in a way that made its reading a unified whole, building its message

page by page with purpose and increasing significance in order to have ordinary people relate to its diverse set of lessons. I wrote this book to synthesize what I have learned into an understandable and, I hope, inspiring context for my family and others.

3. I surely wish that I had learned all of this in Sunday School, in Religious School or earlier in my life. It would have made my life even more grand than it is already!

There is a great deal to think about and relate to in these pages, so please do not feel compelled to read this book in one sitting. Additionally, the first three chapters concentrate on discussions of science and religion, while the remainder of the book is more focused upon the biblical role of Man. The two sections can be read separately; however, together the significance of the Book of Genesis truly resonates.

I hope that you enjoy taking this journey with me. Even more, I hope that reading this book starts you on your own journey. What I explore here is limited to Genesis and my own impressions and interpretations. As you develop your own thoughts, they may take you to other books of the bible, essays on religion, scientific journals, historical works, long talks with friends about spiritual guidance or in any number of other directions that all have the potential to enrich your life.

CHAPTER 1

WELCOME TO THE BOOK OF GENESIS-HOW DOES JUDAISM MAKE IT THIS FAR?

NOTE: If you have not read the Preface and Introduction, please take a few moments to do so. They delineate the perspective from which this book is written.

One of my first observations was that the Bible teaches us in a variety of ways. It most often explores an issue directly. However, the Bible also uses subtle techniques, couching the lesson sometimes in the power of a single word, sometimes within a scene that initially seems out of context and sometimes by leaving part of an episode unaddressed or unresolved, forcing the reader to play an active role in gleaning whatever significance is implied.

Even before we read the first words of Genesis, we come upon our first challenge: the unlikely existence and survival of Judaism itself!

Judaism was introduced into the ancient Near East—with Mesopotamia to the north and east, Greece to the north and west, Egypt to the south and the Land of Canaan in the midst of it all—at a time when there was, of course, no TV, no modern physics, no modern cosmology, no Discovery Channel, no weather reports, no geology and no talk radio to attempt to explain natural phenomena. There existed only a world full of pagan gods presumed to manifest the people's beliefs about good and evil through the divination of nature and its parts. For example, the sun, the moon, volcanoes, thunder and various animals such as rabbits and field mice were seen as independent deities,[1] while humankind was directed toward the goal of attaining immortality.[2] There was little reason for the people living at that time to think that they should believe differently.

Into the midst of this cauldron of beliefs entered the people of the Hebrew GOD who challenged, at their most basic level, the spiritual ideas of the known civilized world. These Hebrews asserted that there was only one GOD, that GOD was everywhere even though no one could see or touch Him and He had no earthly manifestation, that GOD was simultaneously ever-present and unknowable, and that GOD was interested in neither power nor human immortality but, rather, in justice, mercy and morality.

Just imagine what the neighbors must have thought!

To place this in context, we know today that even with modern, sophisticated scientific research and mass media marketing, we cannot convince people to stop smoking. We start to educate children early in life, we bombard the airwaves with health warnings, we try to make it un-cool and un-sexy to smoke *ad infinitum.* And still people smoke. Similarly, modern Man has the intelligence to understand that failing to buckle a seat belt is very dangerous and the mass media can get the message out. Yet, even in the face of overwhelming information many people continue these dangerous practices. In ancient times, when the "Internet" consisted of writings on

---

1   Thomas Cahill, *The Gifts of the Jews* (New York: Nan A. Talese/Anchor Books div. Random House, 1998), 16.
2   Cahill, *The Gifts of the Jews;* Gary Rendsburg, *The Book of Genesis, Vols. 1 and 2* (Chantilly, VA: The Teaching Company Lectures, 2006).

papyrus and parchment and the news of the day was relayed by businessmen traveling in caravans along established trade routes, information traveled slowly and beliefs were even harder to change than they are today.

Try to imagine what the people of the ancient Near East, without the pervasive knowledge and sophisticated media that can so effectively disseminate information and different points of view, must have thought when they came upon the Hebrews and their strange idea of GOD. The fact that Judaism, with its entirely different outlook on the essence of everything spiritual including the revolutionary idea of monotheism, even made it through a first exposure seems impossible on its face.[3] You'd expect such a religion to have been declared laughable by the people in the ancient Near East and antithetical to their core beliefs. It should have been derided, stopped and banished. It confronted the belief systems that the people held as sacred and that, therefore, directly threatened their deep-seated truths. Since there was nothing in their life experience to prepare them for the possibility that these presumed truths could be challenged, you would think that these people would have had the Hebrews and their new ideas quickly drummed out of existence.

Despite unimaginable odds, the Jewish people and their concept of GOD do still exist. The Jews had neither political nor military power to use against the status quo. They simply had a Torah where God learns what makes human beings function and where people wrestle to understand the complete essence of GOD.

---

3   Rendsburg, *The Book of Genesis.*

To further exemplify the differences between the Hebrew and pagan viewpoints, we need go no farther than the great literature of antiquity. The Homeric stories and Gilgamesh Epic de-emphasize ethical and moral directions, choosing instead to have humans try to approach divinity through the acquisition of personal immortality.[4]

In contrast, the Torah promotes not individual aggrandizement or human immorality but, rather, Man's attempts to relate to the sovereignty of GOD. It differs from Greek, Sumerian, Babylonian and other Mesopotamian tomes in another way as well. The Torah, and specifically the Book of Genesis, should be read as a whole in order to comprehend the full depth of its relevance for humankind. As we will see, each part builds upon the prior section; none stands alone as an independent entity. The Torah is structured to form an ever-strengthening series of layers of morality and ethics.

What is truly inspiring is that the people responsible for many of the concepts underpinning the legal, moral and civil codes of modern western cultures arose from this seemingly impossible set of initial circumstances. This is beyond conventional wisdom!

So, if you believe in miracles, the fact that the Hebrews made it this far could surely be one!

4   *McGill's Medical Guide*, revised ed. (Salem Press, 1998); Michael Mc-Goodwin, "Epic of Gilgamesh, Summary" (prepared 2001 and revised 2006), at http://mcgoodwin.net/pages/otherbooks/gilgamesh/html; Rendsburg, *The Book of Genesis;* Rabbi Joseph Telushkin, *Biblical Literacy* (New York: Harper Collins, 1997), 16.

CHAPTER 2
HOW THE PHYSICS OF THE BIG BANG LEAVES ROOM FOR GOD

"In the beginning of GOD's creating the skies and the earth..." (Gen. 1.1).

These first 11 words of the Bible start us on our journey into awesomeness. Let's look carefully at what is being said. The Bible begins with the creation of the skies and the earth, not the creation of the universe.[1] Since the "items and materials" required to produce the skies and earth were seemingly already in existence, it is appropriate at this point to ask: what came before the earth's creation? As we search for the answer, we will already begin to understand that science and Judaism are not incompatible. Let's start with a review of some of the physics of the universe and the science of cosmology. (For a more detailed explanation of the science and relevance of the Big Bang, please turn to the first Addendum at the back of the book.)

Prior to the moment of the Big Bang there existed conditions in which our currently known understanding of physics breaks down. If you were to start counting backwards from today for 13.7 billion years, which is the latest estimate of the age of the universe, you would arrive at the inception point of the Big Bang.[2] This point is called the "Singularity." Just an instant before that, you would find a void—formless and filled only with virtual particles and potentiality rather than mass, suspension rather than time.

1   Richard Elliott Friedman, *Commentary on the Torah* (San Francisco: Harper Collins Publishers, 2003); Matt, *God and the Big Bang.*
2   Bryson, *A Short History of Nearly Everything*; Tyson and Goldsmith, Origins: *Fourteen Billions Years of Cosmic Evolution.*

At the moment of the Singularity, physics declares that time in our universe begins.[3] The laws of physics state that at the very inception of the Big Bang there must have been infinite mass (density), infinite gravity, infinite heat, infinite curvature of space-time and infinite energy.[4] Yet all of this is clearly not possible, according to famed physicist Stephen Hawking, "as predictability would breakdown. In other words, the entire universe was squashed into a single point with zero size."[5] From this unimaginable state, some unknown force put into motion a process that began to expand and change the Singularity in such a way that resulted in the entirety of the vast universe that we know today—from the pervasive background microwaves to the enormous gas clouds that condensed into the supernovas that produced early elements such as carbon and iron plus the stars, the planets of every galaxy including our small solar system, our earth, its life and all of the rest of the cosmos. Modern physicists are able to trace every single atom of every single thing back to this instant and, again, since the mathematics of the laws of physics break down at the instant of the Singularity, our time and history must have their beginnings at this very moment.

There are currently several theories contributing to a "Grand Unified Theory," the goal of which is to explain the origin of every physical thing. String theory, Membrane Theory and Multiverse Theory are a few of these, but remember that they are just theories. What

---

3   Paul Davies, *About Time* (New York: Simon & Schuster, 1995); Hawking, *A Brief History of Time*, 69.
4   Hawking, 68-69.
5   Ibid.

is not theoretical is our Universe's oneness, which is grounded in scientific reality.[6] As we can now understand, the physical order of our cosmos is universal and we, humankind, are part of it.

"Every atom you possess has almost certainly passed through several stars and been part of millions of organisms on its way to becoming you," Bryson says in his book.[7] Neil de Grasse Tyson and Donald Goldsmith, authors of *Origins: 14 Billion Years of Cosmic Evolution*, the companion book to the heralded NOVA miniseries, put it this way: "Every one of our body's atoms is traceable to the Big Bang and to the thermonuclear furnaces within high-mass stars. We are not simply in the Universe, we are part of it."[8]

For another example of cosmic oneness, lets take a look at Einstein's masterful equation, $E=MC^2$, which states that matter is not distinct from energy, but is energy that has temporarily assumed a particular form.[9] Matter is tangible; matter and energy are different states of the same thing. Furthermore, while matter can be transformed, it cannot be eliminated[10] and, since the operative symbol of this equation is *equals*, we can see an elegant display of how one of physics' most marvelous formulas demonstrates the mutuality between all things.

Science's understanding of the ultimate origin of

---

6   Matt, 35.

7   Bryson, 134.

8   Neil deGrasse Tyson and Donald Goldsmith, *Origins: Fourteen Billion Years of Cosmic Evolution* (New York: W.W. Norton and Co., 2004), 29.

9   "E=MC2 Explained," at http://www.worsleyschool.net/science/files/emc2/emc2.html.

10   Bryson, 100.

the universe falls short, with no consensus emerging.[11] Thus, the force responsible for the resultant vastness that completely encompasses everything in our universe, including us, is still unknown.

The specific details of what happened before the Big Bang are not as significant as the fact that whatever it was that sparked the initiation of the Big Bang is responsible for everything that we comprehend, sense, see and are today—the order of mathematics, light (photons), gravity, mass, strong and weak forces and everything else in our universe. The enormous power of the supernova that produced fundamental elements such as carbon and iron and propelled them throughout the entire universe, along with the energy required to complete these tasks, exists from whatever it was that initiated the Big Bang. "The gold in our jewelry and the uranium in our nuclear reactors are both remnants of the Super Novas that occurred before our Solar System was born," writes Stephen Hawking.[12] The potentiality of the Singularity is manifested in the actuality of the universal oneness that now exists.[13]

Thus, science has confirmed for me that we are stardust, that the entirety of this continuum may not be knowable to Man and, therefore, the force that started it all is ultimately unknowable to me. This unknowable "force" I choose to call GOD.

As we will see throughout this book, science does not preclude the existence of GOD. Science confirms that everything in the universe has the same origin, which allows

---

11   Matt, 28.
12   Hawking, 83.
13   Bryson, *A Short History of Nearly Everything.*

us to accept this fact and move the discussion forward. We can now seek to identify the aspects of God to which Man is capable of relating while simultaneously seeking to comprehend the awesomeness of the aspects of GOD that are not totally accessible or knowable to Man.

Science demonstrates that mankind is part of the grand oneness of the universe, and the Hebrew religion directs us to try to relate to this oneness while acknowledging that not all aspects of this oneness will be knowable to Man. Thus, I find no inherent contradiction in stating that the search for the amalgam of everything physical and spiritual is a search for what I call GOD.

So those first 11 words of the Book of Genesis—"*In the beginning of GOD's creating the skies and the earth*" (Gen. 1:1)—may indicate that GOD is simply using materials that existed as a result of the Big Bang. From virtual nothingness, from an indescribable void, has come the current uncountable vastness; from the unanswerable supported by the unknowable comes the universe we inhabit. Before we move on, let's take a moment to get a real sense of the grandeur and grand scale of the universe. What is the magnitude of the physical "everything" that we now call the universe?

Cosmologists tell us that there are approximately 100 billion galaxies, each containing approximately 100 billion stars. This does not include planets, moons, asteroids, comets or any other objects.[14] Just for fun, note that the current estimate of the number of neurons in the human brain also is approximately 100 billion.[15]

14   Bryson, *A Short History of Nearly Everything;* Robert Caldwell in *Scientific American* (Sept. 24, 2007).
15   *McGill's Medical Guide; Scientific American Book of the Brain* (New York:

Quite a coincidence!

This feels like a lot of "stuff" in our universe!

Try to imagine it—100 billion galaxies, each having 100 billion stars plus the almost incalculable number of miscellaneous objects. In addition, there is the untold cumulative mass of the giant Black Holes that are believed to be at the center of each of these galaxies.[16] Still, it turns out that all of this is not really very much when compared with the total amount of mass that physicists calculate must exist in order to corroborate the math that rules all of the laws of physics.

What does this really mean? Hold on to your hat!

The sum of all of the mass listed above is only 4% of the total mass that physicists calculate the universe must contain! That's correct: just 4%. We simply cannot find the rest of it. We know that much more mass must be there because of the observable effects this missing mass has on an individual galaxy or the universe as a whole; it's just that we cannot "see" it![17] The calculation is complicated by particles such as the neutrino, trillions and trillions of which are constantly passing through our planet and everything on it, including you and me![18]

Doing the math, this means that 96% of all mass in existence is currently unaccounted for.[19] This enormous and necessary amount of mass, both the known and the unknown, resulted from the inception of the Big Bang

The Lyons Press, 1999).

16   Tyson and Goldsmith, *Origins: Fourteen Billion Years of Cosmic Evolution.*
17   Ibid.
18   Bryson, 194.
19   Tyson and Goldsmith, 61.

during a period of undefined physics before time began. A cauldron of potential: a void filled with virtual existence. Our universe came from seeming nothingness to a volume of mass larger than we can comprehend!

Now as the skies and the earth are about to be made known to us in the Book of Genesis, let's consider the probability of the earth's even making it through its period of formation. We know that the earth underwent an onslaught of changes of epic proportions in order to appear as it is today, nearly four billion years later. All sorts of atmospheric and environmental changes took place—lands becoming oceans; lakes becoming lands; continents adrift fueled by continental tectonics, with ice cover and then melting during four ice ages; and a constant bombardment by countless meteors. All of this characterized our planet's "birth."

The earth's amazing survival story hinges upon a marvelous combination of conditions that include:

1. If the moon had not remained at just the correct distance from the earth for the earth's axis to tilt at 23 degrees, the tides and trade winds critical to our survival would simply not exist.[20]

2. Without the earth's molten core, we would not have the electromagnetic field in place that so efficiently shields us from the solar and cosmic rays and their harmful effects.[21]

3. Without Saturn and Jupiter being precisely where they are in relationship to earth and at their current rel-

---

20   Bryson, 248.
21   Ibid.

ative sizes, we long ago would have been blasted out of existence, because these planets have acted as vacuum cleaners for us as they either "catch" or deflect potentially catastrophic meteors and asteroids that may have hit earth. And how many asteroids might there be in "our" space?

As Steven Ostro of the Jet Propulsion Laboratory explains it, "Suppose there was a button you could push and you could light up all the earth-crossing asteroids larger than about ten meters. There would be over 100 million of these objects in the sky...all of which are capable of colliding with the earth, and all of which are moving on slightly different courses through the sky at slightly different rates. It would be deeply unnerving."[22]

4. Without the extremely unusual properties of water, life as we know it may never have been possible. When bodies of water freeze, they freeze from the top down, leaving unfrozen water below the ice top, which allows life to continue below. Other fluids such as silicone freeze completely from top to bottom and bottom to top, leaving no room for life to continue. If water had assumed these typical liquid properties, we would have had one solid block of cold during the Ice Ages, with nowhere for life to make its home![23]

5. If the wavelengths emanating from the Big Bang had been just .0001 shorter, gravity would have taken control far earlier in the universe's history. A total contraction of everything would have occurred, leaving no universe as we now know it. If the wave-

---

22   Bryson, 194.
23   Rabbi Bradley Savit Artson, *The Bedside Torah* (New York: McGraw-Hill, 2001; Bryson, 271; Tyson and Goldsmith, 255.

lengths had been just .0001 longer, on the other hand, gravity would not have had the appropriate attraction, permitting everything to continue to expand forever. No contraction of particles, again no universe as we now know it. Writes Bryson, "What is extraordinary from our point of view is how well it turned out for us. If the universe had formed just a tiny bit differently—if gravity were fractionally stronger or weaker, if the expansion had proceeded just a little more slowly, then there might never have been stable elements to make you and me and the ground we stand on. Had gravity been a trifle stronger, the universe itself might have collapsed like a badly erected tent, without precisely the right values to give it the right dimensions and density and component parts. Had it been weaker, however, nothing would have coalesced. The universe would have remained forever a dull scattered void."[24]

These are just a very, very few of the criteria upon which the extremely fragile existence of the earth is based. Since there are many more examples, I will leave it to the reader to think about the probability or improbability of our being here today after 13.7 billion years of such unimaginable change. What we should be pondering is: what are the implications of these probabilities regarding the earth's existing as we know it today? But, while modern science provides us with an immense amount of knowledge regarding the *how's* and *what's* of our universe in cracking the genetic code and the cosmic code,[25] science is much less adept at helping us to understand the *why's*.

24   Bryson, 15.
25   Matt, 62.

Why did the Big Bang occur in the first place?

Why did the sun just "turn on"?

Why did the planets build themselves from gas and dust? "In simple summary, we can state that no good explanation exists of how the planets 'began' to build themselves from gas and dust," write Tyson and Goldsmith. "The beginnings of planet building pose a remarkably intractable problem."[26]

Why did the giant Black Holes, at the center of each galaxy, hyper-condense into these cosmic magnets?

Why is the order of mathematics universal?

Why are the galaxies racing away from each other at an accelerating rate?[27]

Why are the outer stars in a galaxy held fast to their positions and not flung out into space?[28]

Why does a group of atoms conjure up within our brain emotions such as anticipation, anxiety, wonder and awe?

Why did life start on earth? This is the most interesting one of all because it can be read as:

WHY did life start on earth?

Why did LIFE start on earth?

---

26   Tyson and Goldsmith, 184.
27   Bryson, 171; *Science Today,* November 15, 2002; Tyson and Goldsmith, *Origins: Fourteen Billion Years of Cosmic Evolution.*
28   Bryson, 171; Caldwell, *Scientific American.*

Why did life START on earth? or

Why did life start on EARTH?

Other observers have found, too, that the *why's* are not obvious. "Exactly how no-one knows," writes Matt.[29] "The origin of life on Earth remains locked in murky uncertainty," agree Tyson and Goldsmith.[30] Concludes Bryson, "The upshot of all this is that we live in a universe whose age we cannot quite compute, surrounded by stars whose distances we don't altogether know, filled with matter we can't identify, operating in conformance with physical laws whose properties we don't truly understand."[31]

Thus knowing that science is ensconced in the incomprehensible conundrums of the *why's* allows me to acknowledge something larger and more grand than I can know. Science marvelously identifies the physical oneness of the universe and magnificently demonstrates our vast knowledge of the *how's* and *what's* of our surroundings, making it possible to retrace the full physical history of this universe. In so doing, science sets a wonderful perspective from which Man may regard the awesomeness of our universe while neither disavowing nor disproving the possibility of GOD.

Therefore, the combination of the *why's* of spirituality and the *how's* and *what's* of physical oneness is what I choose to call GOD. I understand that my journey will be a melding of the knowledge, comprehension and study of what *is* knowable to humankind with those areas that are *beyond* Man's full understanding.

---

29   Matt, *God and the Big Bang.*
30   Tyson and Goldsmith, 235.
31   Bryson, 172.

Now that we can understand that science and religion are not incompatible, we are able to comprehend that, yes, there is "room" for GOD.

Now we can go forward.

CHAPTER 3
"LET THERE BE LIGHT!"- GOD AND THE PHOTON

"When the earth had been shapeless and formless..." (Gen. 1:2).

As GOD was at "the beginning of creating the skies and the earth," no earth yet existed; Genesis describes the earth only as "shapeless and formless." From this void a planet was created as it coalesced from the matter made possible by the Big Bang, and this planet was lovingly called "earth." (Take a moment to review several paragraphs in the preceding chapter. Do the words seem familiar? Do you find scientific descriptions of the unknown sounding somewhat similar to the spiritual descriptions?)

"God said, and 'let there be light', and there was light. And God saw the light and it was good" (Gen. 1:3). Biblical scholars view GOD's Light as the bringing of Good to the earth.[1]

As we have learned from the science of cosmology (See Addendum 1), for the first 380,000 years after the Big Bang, the universe was merely a particle soup so dense that even the energy created from the primordial collisions of electrons and positrons, particles and anti-particles, matter and anti-matter could not escape.[2] As time went on and the universe moved past 380,000 years of age, the temperature cooled slightly and the pace of these annihilations slowed, allowing the energy that resulted from this process to finally escape out into space.

---

1   Rendsburg, *The Book of Genesis.*
2   Tyson and Goldsmith, 43.

In what form did this energy leave the "particle soup"? It left in the form of photons which are also called "packets" of light! As we read GOD's very first words from Genesis—"Let there be light"—we realize just how wondrous and exhilarating these words truly are.

It was the light from the very inception of the Big Bang that preceded everything else to announce the dawn of the universe, kind of like GOD's calling card. Way, way before any planet would or could form, this light let it be known that something was coming. The darkness of the void was eliminated, the potential of things to come illuminated.

How marvelous it is that the Bible says that GOD's first words are, "Let there be light," and that science has proven that the photon of light arrived before anything else to show the way. But is this intended to absolutely define the essence of GOD? No.

Again, I believe that the essence of GOD is beyond the capacity of Man to completely comprehend. However, as we see from the paragraphs above, there is a profound symmetry between the first "spoken" words of the Bible, "Let there be light," and the currently accepted scientific theories of the birth of our physical universe as announced by the photon release. The two realms are clearly not incompatible.

We therefore start with an indefinable dimension of everything that I call GOD. At the beginning of the Book of Genesis, order is fashioned from chaos, and the results of inevitable entropy are actually the seeds of what will come next. Man begins to encounter both an unknowable spiritual entity and the unanswerable *why's* of science. Therefore, Man becomes aware that the unknown permeates his existence. This pervasive unknown I choose to call GOD.

As we now begin to move into the text of the Book of Genesis, it is important to make this point: in order for a human of limited capabilities to relate to a GOD of everything, we should recognize that the Bible is available to help us understand the spiritual essence of an unknowable GOD and, equally as important, to also become familiar with those moral and inspirational parts that *are* relatable to Man. These relatable aspects, which I call God, are garnered from the relationships of GOD to Man and Man to Man.

Early in my studies I realized that it would be necessary for Man to try to comprehend both GOD and God in order for us to have any chance of reaching our potential. The Torah is ready to guide humankind and should be understood as the "place" where a limitless GOD and a Man of limitations come together and find ways to meaningfully relate, in order that Man may pursue a path toward holiness.

GOD is the source of both our Light and the cosmological light, simultaneously a beacon of guidance and illumination.

So GOD now continues His work, and the Torah begins!

CHAPTER 4
DEMYTHOLOGIZING NATURE

GOD created nature and is not just *of* nature. Pointed out by author Gary Rendsburgy in *The Book of Genesis,*[1] this is a fundamental distinction between the Torah and the beliefs held throughout the ancient Near East. The Bible drives home this point by presenting nature as demythologized and taking its rightful place among GOD's marvelous wonders.

In going to great lengths to demonstrate that the Hebrew GOD is not limited to simply being one of nature's many gods, the Torah contradicts the hallmark religious beliefs of the ancient Near East cultures. The Hebrew GOD would not, could not and must not be confused with the pagan gods of nature such as the sun, the moon, field mice and mountains. The Book of Genesis leaves room for absolutely no confusion here: the Hebrew GOD created the conditions for nature and is not *of* nature. GOD made possible the waters, volcanoes and thunder and is not merely one of them or similar to them.

In its most striking example of the importance of this understanding, the Bible states:

"GOD made two big lights—the bigger for the regulation of the day and the smaller light for the regulation of the night" (Gen. 1:16).

It is obvious that this refers to the sun and the moon. However, these particular words, "sun" and "moon," are

---

1   Rendsburg, The Book of Genesis.

intentionally not used in the biblical text as there is to be no mistaking a sun or moon being a GOD in this Book![2]

2   Ibid.

CHAPTER 5
"IN GOD'S IMAGE"

"And GOD created the human in His image. He created the Human in the image of GOD; He created male and female" (Gen. 1:27).

Here we have GOD setting the stage for the rest of our journey.

The phrase "in GOD's Image" is one of the most inspirational, yet sublimely difficult, phrases in the Torah. It inspires without being defined. Even without specificity, we feel its essence. It implies, says Richard Friedman in his *Commentary on the Torah*, "that humans are understood here to share in the divine."[1]

This phrase has been interpreted throughout the ages either as the encapsulation of the sense of humility, morality and wonder given to Man by GOD, or as a demonstration of Man's hubris and conceit. Man should be humbled by the implications, not become haughty at having been selected to represent "specialness."

Being created in GOD's image brings Man an awareness of his potential godliness; it is this awareness that separates us from the remainder of GOD's creations. But differentiation and awareness, though wonderful, should not be enough for humankind to become complacent.

I choose to regard the "in GOD's image" description as the initiation of Man's responsibility to recognize right from wrong, justice from injustice and good from

---

1   Friedman, 12.

bad. It is a statement of our requirement to try to act in a godlike manner toward other human beings who also have been created in GOD's image. "Try" is the operative word here, as we will see that the extraordinary expectations implicit in this phrase can be a burden—and a heavy one, indeed.

In the first chapters of the Book of Genesis, GOD starts our journey on a wholly positive note; at the end of most of the days of creation, GOD affirms that what was done that day "was good." Much of the remainder of the Torah has GOD adding layers of clarification and meaning to an enriched definition of "in GOD's image," as GOD wishes to have Man incorporate this image of goodness into his daily life and apply it to our interaction with each other. Unfortunately, Man will need some help living up to GOD's expectations. "Though I am created in the Image of GOD, we relate to each other as other," Daniel Matt notes.[2]

Being "in GOD's image" elevates Man only if he exhibits godlike behavior and sensitivities and, as the ultimate challenge to humankind, it is the standard upon which I believe the entire Torah is based. GOD wants Man to act in harmony with justice, righteousness and holiness and begins to place this burden squarely upon Man's shoulders.

Are human shoulders broad enough to handle the pressure? Let's see!

---

2 Matt, 48.

CHAPTER 6
"THE SEVENTH DAY" - HOLINESS

"And on the seventh day GOD finished His work that He had done, and GOD ceased on the seventh day from all of His work that He had done. And GOD blessed the seventh day and made it holy" (Gen. 2:2-3).

Here, at the Torah's very beginning, is where one of Judaism's most fundamental contributions is elucidated as we are told that holiness is not to be associated with a specific place or a specific thing but, rather, holiness is associated with time.[1] While Man may appreciate a thing or a place, holiness is to found in time, in space.

Thus, Man is to take time to reflect, to study, to love and to incorporate the sustainable message inherent in the phrase "and it was good" into his daily life. We are to take the time necessary to carve the message of GOD's goodness into our consciousness and to do so in any place and at any time; in fact, we will see the concepts of "any time" and "any place" repeated throughout Genesis.

GOD spent six days creating wonders far grander than anything that we are capable of making ourselves. He fashioned order out of chaos and organized it into a miraculous world. Yet, what GOD declares as holy is not the magnificence of what He made during these six days. It is the following day, the day itself upon which GOD rested, that He declares as holy.

When we take time to rest on the Sabbath, we are recreating GOD's notion of holiness. As we use this day

---

1   Rendsburg, *The Book of Genesis.*

to let order reemerge from the chaos of our daily lives, we can then take time to reconsider what is truly important in our existence. We reconnect with what "in GOD's Image" might mean as we try to avoid becoming so busy that we are unable to take stock of ourselves and the oneness of everything holy.

As we explore this concept—that holiness is found in time within space, not at a place in time—we see why holiness cannot be found in the physical entities of our world. Just as holiness is not measured by how big the planets are that GOD made, neither is it implied by the fleet of cars that we own, the square-footage of our home, the size of our bank account or the number of camels we have in the stable out back. Holiness is achieved through taking the time for the reflection, conscious awareness and thoughtfulness that can lead to justice, harmony and peace. A.E. Herschel wrote that "the special significance of the concept of the Sabbath is that it means the sanctification of time," according to Friedman.[2]

How clearly is this message perceived in the Jewish heart? One of the very first lessons that we are expected to take from the Torah is directed by something we cannot physically touch, yet it instructs us to take the time required to become grounded in those things that touch us and are truly meaningful to others. Holiness is found by establishing a separate time for reflection, away from the everyday requirements of life. It is associated with a time to reflect upon the preciousness of our existence; it is an opportunity to become spiritually in touch.

---

2   Friedman, 15.

What a wonderful message is this!

We know that today's physicists are working feverishly to find the Grand Unified Theory of everything physical, to join General Relativity and Quantum Mechanics in the perfect marriage that will reveal one set of laws for the entire physical universe. I like to think of the study of Torah as our search for the Grand Theory of those things that are holy, the unification of godliness and human purpose leading toward spiritual oneness.

Whereas, just as in physics, Judaism acknowledges an unknown force extant that we do not understand and perhaps never will, the possibility of increased perception does exist; therefore, the search is important in Judaism. Also, just as in physics, the more we ask and learn, the more questions seem to arise. The asking, therefore, makes the searching ever more interesting, exciting and necessary.[3]

As the Torah continues, we are increasingly introduced to the aspects of God to which we humans can relate. We begin to experience an ebb and flow of GOD and God, with the Torah seamlessly exposing us to the unknowable essence of GOD while weaving in those aspects of a God to which we are capable of relating. One of the brilliant insights of Judaism is this duality, which enables us to approach holiness as we proceed on our march toward living in GOD's image.

These three words, "in GOD's image," have even greater impact when you consider that GOD watches us struggle to figure out the true meaning of the phrase;

---

3   Rendsburg, *The Book of Genesis.*

GOD becomes the guide and God the road map: one the goal and one the means to achieving the goal. GOD sets the standards, and God identifies the actions required to meet them.

Thus, we understand that the Torah intends for Man to take time for reflection, for acknowledging the world around him, for acting in GOD's image and for recognizing holiness.

CHAPTER 7
CREATION EPISODE TWO - IT'S MAN'S WORLD

Just when you may think that things are settling down—the universe has been formed, the earth seems to be in order, human beings are in place and the standard to live in holiness had been set—what does the Torah do now? As so often happens in the Bible, the moment at which things seem to be going in one direction is exactly when the Torah challenges us and takes us in another direction.

The next portion of The Book of Genesis is a perfect example as it restructures and introduces us to a second creation episode! This episode tells us of a God that is now interested in a relationship with humankind, a God less concerned with the overwhelming realm of the universal. He is ready to directly relate to Man.[1] Here the Torah sets a clear exclamation point that marks the formal introduction of what the phrase "in GOD's image" should mean to Man. In so doing, the Torah makes it eminently clear that Man will have the tools and guidance required to live in holiness—*if* man so desires!

Starting at Genesis 2:4, this second creation episode moves from the universal cosmological environment found in creation episode one to focus more heavily upon the relationship of God to Man.[2] Reacting to the challenges of living in GOD's image exposes Man's limitations, and we start to sense just how important the Torah may become in assisting Man on his quest.

---

1   Ibid.
2   Ibid.

In subsequent chapters, Mankind will be guided by a formal covenantal structure in order to assist in achieving a life of godliness. Since Man will have a difficult time learning these lessons, a road map will begin to be laid out for him. The key question is: will Man choose to follow it?

The wonderful, poignant and powerful word throughout the remainder of Genesis is "chooses." We will see as we continue through the Book of Genesis, humans will be allowed to determine how their role will be defined as they can *choose* whether to maximize their potential for goodness. It is this ability to choose that makes the Torah in general, and the underlying messages of the Book of Genesis in particular, so ensconced in optimism. Yes, Man might need some help, but the attainment of godliness and holiness *are* within his grasp. We will see that the GOD of the Hebrews will always leave the door open to increased godliness for anyone who chooses to step through its portals. It is through Man's "choosing" that the world can always become a better place; therefore, the effects of increased goodness may be just around the corner.

Essentially, Man and God enter into a relationship that increasingly moves Mankind toward living in GOD's image. Thus, the Book of Genesis should be taken as a whole and not viewed simply as individual vignettes or morality plays. By observing Man's behavior over the duration of the entire saga, we can gauge his willingness at times, and reluctance at other times, to follow God's road map. The extent of Man's ongoing commitment to this road map will serve as the barometer by which we will judge how well Man is progressing.

This all begins to transpire with the second creation episode. This creation story is anthropocentric, not cosmological,[3] with Man positioned at the center of the action.

For example, the order of creation is turned inside out. In this second version, the solid earth is created first and the heavenly sky second, rather than the skies coming first as in the first version. Next comes Man, then vegetation and finally the animals. There is no mention at all in the second creation episode of the sun or the moon.

The language also is more human-centered. The Torah specifies that GOD:

- *forms* the earth.

- *fashions* man.

- *plants* the garden.

- *fashions* the animals.

- *constructs* woman.[4]

GOD's actions are communicated in ways to which Man can relate; the words could as easily be describing Man's own actions. We also find in this creation version a "touchable" earth, not at all like the ethereal skies of creation episode one. There is no "thinking" this earth into existence; it is a world that revolves around Man

3   Ibid.
4   Ibid.

and shows a "hands-on" God who positions Man first among all created things.

Once more, the genius of the Bible unfolds, as throughout these pages the cosmological/universal GOD and the anthropocentric/relatable God are juxtaposed. This juxtaposition is initiated at the beginning of the Book of Genesis in order to have Man clearly know how necessary it is for him to try to comprehend the different aspects of GOD, to realize that GOD is manifestly beyond human conception and to simultaneously understand that Limited Man is also capable of directly and meaningfully "knowing" the aspects of a relatable God. Both aspects are required and, so, both are presented. Together they form a unified foundation upon which Mankind may conceive of the full holiness of GOD.

We now have the groundwork set. We know that there is a potential to be achieved and a challenge to be addressed. So how does the Torah help Man to live in GOD's image?

Enter the Tree of Knowledge of Good and Bad in the Garden of Eden.[5]

In the Garden, humans' ability to make choices forever changes their relationship with both GOD and God and presages the dramatic impact that choice will have upon all future relationships.

---

5   Friedman, 17. Friedman prefers the Hebrew translation of "Good and Bad" over the more widely accepted "Good and Evil," since "evil" suggests only moral knowledge. He writes, "The Hebrew translation has a much wider range of meaning than that."

CHAPTER 8
THE GARDEN OF EDEN- MAN'S PLACE IN THE PLAN

"And YHWH [God] planted a garden in Eden at the East, and He set the human whom He had fashioned there. And YHWH caused every tree that was pleasant to the sight and good for eating to grow on the ground, and the Tree of Life within the garden, and the Tree of Knowledge of Good and Bad" (Gen. 2:8-9).

"And YHWH commanded the human saying, 'you may eat from every tree of the garden. But from the Tree of Knowledge of Good and Bad you shall not eat'" (Gen. 2:16-17).

What is this Tree of Knowledge of Good and Bad, and what will be its impact upon humanity? Let's use a literary example for perspective. When we say, "They came to the meeting young and old," we mean that people of all ages came to the meeting, not just literally the youngest and oldest. We can use the same interpretive style here.[1] The Bible is not just speaking of the specific knowledge of good and bad but is indicating that Man's mind should be open to his potential to acquire all knowledge. Thus, as a result of having eaten the fruit of this tree, humanity becomes aware of the ability to attain good, bad and everything in between. All knowledge is now within Man's grasp. This awareness includes consciousness, beliefs, pride, embarrassment, anxiety, anticipation, apprehension and much more.

Judaism acknowledges an evil inclination, "Yetzer Ha Ra," within Man, but Judaism does not believe in the separate existence of a cosmic evil force. It recognizes that Mankind's sexual urges, ambition, approval-seeking

---

1   Rendsburg, *The Book of Genesis*

and the like are part of human existence that should be channeled and oriented in the direction of GOD's will.[2]

As a result of eating this fruit, Man comes face-to-face with all levels of good and bad. It will become increasingly apparent that the bad will have to be confronted with actions and refuted, and the good taken to heart and reinforced, by a different set of actions. This establishes another key tenet of Judaism: actions are required to approach godliness! Words and thoughts are not sufficient. In the words of Moses Cordovero, "The essence of the Divine Image is action."[3] It takes a single action—eating a piece of fruit—for Man to become aware of the potential of good and bad, but it will take a lifetime of actions to dispel the bad and reinforce the good.

Jews are to be a people of doing, not just assuming. Actions, in fact, will become more important than words. Study must lead to action and transformative behavior.[4]

The act of eating from the tree provides Man with increased cognition, which separates us from the animals and places us "closer" to GOD. What are the additional implications of this action? As humanity takes a small step on its unique path of maturation, we begin to know of our potential for expanding knowledge, reason, fear, wonder, disappointment, shame, awe and every other thought and emotion.

Yes, Man is exiled from the Garden, leaving behind the influence of the oneness of GOD's Eden and feeling

2   Dr. Elliot Lefkovitz, personal notes (2008).
3   Matt, 54.
4   Lefkovitz, personal notes.

estranged from Eden as he finds himself in his new surroundings outside the garden. Still, Man remains part
of the created universe, however obscured his sublime
oneness with GOD has become. Is it possible that Mankind gains something more as a result of his new capacity for knowledge?

Adam and Eve are told directly by GOD not to eat
of the tree. They do so anyway. What other purpose
can this serve? Through this act of eating the fruit,
Mankind establishes and secures the existence of his
free will![5] Though their actions remove them from the
communion with GOD's complete oneness in Eden,
Adam and Eve demonstrate that by exercising their
free will and deciding to eat the forbidden fruit they
gain the ability to search for knowledge. With this new
skill, they are capable of following a path back to the
oneness of Eden—if they so desire. We will learn that
it will take a great deal of effort to use this newly acquired free will to proceed on a quest of regaining the
oneness that was seemingly lost as a result of Man's
acting against GOD's directive.

Now exiled from the Garden, human beings have
not only their newly acquired free will to contend with,
but also their ego. It is within this context that Mankind begins to take the next step. The ego, supported by
free will, transforms the Hebrew Man into a maturing,
independent being who is capable of recognizing the
importance of acquiring knowledge! It is the continuing brilliance of the Torah that it takes this momentous
transformation into account and helps humankind to

---

5   Rendsburg, *The Book of Genesis;* Telushkin, 13.

travel, one step at a time, on a path back to oneness with the GOD of the universe. Each member of the tribe of Man, guided by free will and ego, approaches this trip into exile differently. However, through increased knowledge of the Torah, the everyday actions of any one person might increasingly exemplify Man's living more closely within the essence of godliness.

Biblical Man's challenge was to understand how to use his free will appropriately in order to help all of Mankind establish a more godlike world. When someone is simply handed a trophy or an award without having put forth the required effort, the trophy is meaningless. The Torah's implied message is: "You know of the communion with and consciousness of the 'oneness' of God from being in Eden" So here, at the very start of the road back, GOD seems to be postulating that since Man once knew of the blessedness of Eden, he should be able to master the knowledge required to regain the blessedness which he had seemingly lost.

But is Man up to the challenge?

• Do we decide to use our free will to broaden and deepen our knowledge and wisdom?

• Will we accept responsibility for our actions, as our new awareness requires us to work toward diminishing the bad in the world and helping the good to blossom?

• Can we put forth the effort required to optimize our potential for good?

Man chooses to act contrary to God's directive not to eat of the tree, even though the stated punishment is to "die." Upon eating the fruit Man does not physically perish; however, what does die or evaporate is the ensconced oneness with GOD along with the innocence of Man.[6] Thus, our ultimate challenge from that moment forward is to determine whether we have what it takes to recapture this oneness. Can we accept the responsibilities inherent in our having unlocked this new potential of acquiring knowledge?

The challenge is real. Adam and Eve are "born" into a new reality with no worldly experience or history to guide them. With their innocence lost, the rewritten rulebook exposes them to the temptations of the bad as well as the glory of the good.

Adam and Eve become like newborn babies requiring the guidance of a parent.[7] A good parent knows that the child needs a role model and, therefore, not only verbally establishes rules for the child but, more important, through the parent's own actions establishes standards of morality that the child senses are the real underpinnings of his teachings.

Over the years, the word "Torah" has been defined as "law." A more appropriate translation is "showing the way through teaching." As a child matures, he will make mistakes and test the parent's limits. A parent will have to teach through a well-timed mix of persuading, cajoling, punishing, supporting, defining and rewarding the child. A child innately understands that he can learn from a person to whom he can relate but also knows in his heart that he will never truly understand

---

6   Amy-Jill Levine, *The Old Testament Vols. 1 and 2* (Chantilly, VA: The Teaching Company Lectures, 2001).

7   Telushkin, 13.

everything that makes the parent tick. The child also comes to believe that through increased efforts to communicate, he can increasingly identify those morals that shape the parent's character. The quest is never completely fulfilled, as we cannot fully know ourselves, let alone others, but the parent continually leads the way through teaching.

Again we are impressed by our Torah, a teaching tool that shows GOD acting as the parent and Man, the seeming child, is surely a participant in GOD's plan. Therefore, Man's new challenge is to understand GOD's intentions and demonstrate the ability to learn to exercise free will in a moral manner that can bring increased holiness into the world. Man's separation from the oneness of GOD's Eden places Man and GOD into roles with which they are not familiar. GOD's children have no previous experience upon which to base their new lives, and GOD has never had such contrary behavior to confront.

His new free will allows Man to embark upon a quest of learning to live "in GOD's image." Man must try to conceptualize this image in order to seek it within himself and others, as "in GOD's image" defines the spiritual platform upon which we have stood from before the time of Abraham to this very day!

Now let's go back and see what else is in the garden. No, not the snake, but what about that other tree? The Tree of Life, mentioned only twice in the Book of Genesis, offers Adam and Eve immortality. It is standing there in plain view, yet they do not eat from it or even approach it! Why not?

How does any child respond to an authority figure? What happens when a child is told, "You may eat any cookie except the cookies in that specific cookie jar over there"? As soon as the authority figure's back is turned, of course, the child eats from the forbidden cookie jar.[8]

So, Adam and Eve act like children and defiantly display their free will by going against the directive not to eat from the Tree of Good and Bad. The seemingly innocuous, simple act of eating the forbidden fruit sets all of the people of the Book on an unforeseen yet fabulous journey—the search to overcome human frailty, to regain humility and to recapture holiness before GOD and, in turn, sensitivity and respect toward Mankind.

It is a remarkable choice that Adam and Eve make. They choose to pursue knowledge as Man's most noble quest.[9] By not eating from the Tree of Life, they give up the pursuit of immortality, thus acknowledging immortality as the exclusive realm of GOD.[10] Eating the fruit of the Tree of Knowledge allows humanity to embrace the GOD-given capacity for self-discovery, study, curiosity, learning and morality.

Additionally, free will begins to define our identity, establish our character and mold Man's ego. According to Jewish teachings, man is born morally neutral and must use this free will to pursue the knowledge necessary to regain and retain godliness. Man does not sin *because* Adam and Eve sinned; we sin *as* they sinned[11] and, therefore, we can learn how not to sin. It is up to us

8   Levine, *The Old Testament.*
9   Rendsburg, *The Book of Genesis.*
10   Ibid.
11   Telushkin, 10.

to determine how we will act, today and for all of our tomorrows, since our actions become the yardstick by which Man's moral and spiritual successes and failures will be measured.

This pursuit of knowledge by mortal Man defines his purpose and is at the core of Judaism's cultural, social and religious way of life.[12] Within this context of knowledge, the Torah is there as the "teacher."

Man may choose to act in a godlike manner but, may not choose to be immortal!

Man's quest has now officially begun. Has Man bitten off more than he can chew? Can he regain what he lost? Only through his actions will the tale be told.

If Man does not succeed, he has only himself to blame. Man has made choices; he has chosen to set his destiny into motion. Will he have the character that is necessary to proceed and succeed?

Let's continue with our journey and see how Man progresses.

12   Rendsburg, *The Book of Genesis.*

CHAPTER 9
A LITTLE HISTORICAL CONTEXT

To fully appreciate how much the Hebrews were going against the tide with their search for knowledge rather than a search for immortality, keep in mind that in the ancient near east the rulers of Sumaria, Mesopotamia, Babylonia and Egypt were specifically striving to become not merely godlike but actual ruling gods.

The peoples of these civilizations thought that since the sun, moon, rain and other elements of nature were gods, and since Man was part of nature, that they, too, could become gods.[1] This was evident in the area's classical literature that was widely read and, therefore, well-known throughout the entire region. The heroes of these stories would aim to leave the slavery and servitude of the gods by becoming immortal themselves. They may have thought, "Why not? If a field mouse or the sun can be a god, why can't I?"

But immortality for immortality's sake does not necessarily result in higher moral standards. It merely prolongs existence.

Genesis shows us a GOD of the Hebrews that creates nature but is not a part of nature, a GOD that is greater than nature. The acknowledgment of the awesomeness of GOD has inspired the soul of the Jewish people from antiquity.

As if to underscore the point regarding human mortality, GOD says: *"By the sweat of your nostrils you'll eat bread*

---

1   Ibid.

unto you go back to the ground, because you were taken from it: because you are dust and you'll go back to dust" (Gen. 3:19).

Yes, GOD is immortal, and we are but dust.

Why use the word "dust"? Why not make a poetic reference such as a breeze or drops of the ocean? The word "dust" is truly significant. From dust everything emerges and then again returns. Dust is everywhere, and into dust everything decays. The cycle of life is represented by dust. It is a seemingly insignificant remnant of the very particles that were dispersed throughout the universe from the moment of the Big Bang until now; it is part of all of the "stuff" of the universe. Since we all come from this same stuff, each of us is truly part of the universal oneness.

Furthermore, "dust" reminds us that we are mortal; GOD starts us as dust and ends us as dust.[2] Man's goal is the pursuit of knowledge that leads to holiness, not the pursuit of immortality and the presumptive hubris that would result from our striving to be a god. Dust establishes Man's proper perspective.

Now that the dust has settled...

Let us note that Man chose this gift of knowledge and, from then on, it has been his responsibility to pursue it and use it appropriately! This pursuit of knowledge has defined, and always will define, the Jewish character, identity and life's purpose. Study has defined the Jew.[3]

And to think that all of this started with our deciding to eat a single piece of fruit!

---

2   Artson, *The Bedside Torah.*
3   Rednsburg, *The Book of Genesis.*

CHAPTER 10
"Am I My Brother's Keeper" ? -The Answer is Yes!

Now that Adam and Eve have acquired the ability to gain and use knowledge, their challenge is to use this knowledge in a way that enables Man to regain the complete oneness with GOD that apparently has been lost. So, what do you think happens next? We immediately encounter an episode that shows how difficult this challenge will be for Man.

Cain kills Abel!

One brother kills the other; the younger takes the life of the older. This will not be the last time in Genesis that the younger will dominate an older sibling. It becomes a theme throughout the Book that the seemingly weak overcomes the seemingly more powerful.[1]

The first thing that Man does with his newly found free will is to succumb to one of the most base of human emotions: extreme sibling rivalry![2] For the reader, this episode is intensified because we are not seeing simply man against man; we are watching a brother against a brother.

Interestingly, the Torah does not say specifically why Cain kills Abel. Since the reason is not made clear, what is significant must be the fact itself that Cain chose this specific action to resolve a problem. One brother's slaying of another clearly foreshadows the difficulty that Man will have in trying to live up to the expectations inherent in his newly found capacities and responsibili-

---

1   Ibid.
2   Friedman, 27.

ties. If brothers of the same flesh cannot resolve issues peacefully, how challenging will it be for strangers to secure peace among one another?

Additionally, when GOD asks Cain— *"Where is Abel your brother?"* (Gen. 4:9)— Cain answers in two parts:

1. *"I don't know"* (Gen. 4:9).

2. *"Am I my brother's keeper?"* (Gen. 4:9) The more accurate translation is, "Am I my brother's watchman?"[3]

Man is clearly off to a bad start! With this, the Torah sets a behavioral marker, immediately addressing Cain's words.

First, let's analyze the phrase, *"I don't know."* Cain lies directly to GOD. GOD asked Cain, "Where is your brother?" Yet GOD knows all along that Abel has perished, since He acknowledges that *"your brother's blood is crying to me from the ground"* (Gen. 4:10). Cain has lost touch with the notion that we are all part of the oneness of the universe. He feels that he can differentiate himself by running away from GOD. He is unaware that this is like attempting to run away from himself; there is literally nowhere to hide. As we proceed through our study of Genesis, we will see deception "raise its head" on several occasions.

What about, *"Am I my brother's keeper?"* Here, Cain unknowingly poses the question that will help to establish one of Judaism's prime directives, as the answer is an unalterable "Yes!"

---

3   Friedman, 28.

From the very outset of the Bible, it is established that Judaism will be based upon the concept of caring for each individual, his human rights and his spiritual well-being, and that this caring should result in freedom and justice. From here on, the lessons of the Torah cumulatively answer the question, *"Am I my brother's keeper?"* with a resounding "Yes!"

The Hebrews' collective conscience knows the answer to this rhetorical question even before the question is asked. The world must understand that we are to recognize the dignity of each individual. No distinctions are drawn—rich or poor, tall or short, healthy or ill, powerful or weak. Each person is to be respected. In these early chapters, the Torah establishes this wonderful standard for all of humankind.

Cain's two short responses to GOD's question enable us to further define the essence of our religion as they crystallize the initial concepts of our moral and spiritual potential. This occurs as Man is just beginning to learn to cherish the sanctity of all humanity and underscores Man's responsibility to act in such a way so as to try to recapture goodness by caring for one another.

This emphasis on the sanctity of every individual is the foundation of a long list of moral imperatives that Judaism is proud to have brought to the people of the world.

CHAPTER 11
A SILENT MAN NAMED NOAH

GOD sees that Man is not doing too well.

"YHWH saw that human bad was multiplied in the earth and that in every inclination of their hearts and thoughts was only bad all the day" (Gen. 6:5).

Fortunately, however: *"Noah found favor in YHWH'S eyes: for Noah was a virtuous man. He was unblemished in his generation. Noah walked with God"* (Gen. 6:8-9).

Why is it fortunate that GOD finds favor in Noah? After deciding that Mankind was not living up to His standards, GOD wanted to start over—trash the chess board and reset the game with new pieces. Here we see Man and GOD at the very infancy of their new relationship—one that will, fortunately, grow, change and mature throughout the rest of the Book of Genesis.

Now…how does Noah respond to GOD's decision to destroy everything that is living on the surface of the earth? Ready? Noah says absolutely nothing!

Noah says not one word, makes not one gesture, does not even change his body language to indicate any disagreement with such a harsh decree. Equally puzzling is why GOD would choose to solve this "problem" by taking such drastic action! How about first having a little fireside chat with Man?

What we have here is a clear demonstration that both parties are new to their respective places in this relationship, a relationship that is just beginning.

Noah is like the young child who does not question his parent. Parental authority dictates that the child accepts that the parent is correct and is acting for the child's benefit. The corollary is that GOD doesn't know how to react to these new children who have defied his wishes and have acted in such an unsavory manner. He seems to respond as an inexperienced parent asserting His authority.

So, Noah makes an ark, GOD initiates a flood—and the tension builds!

Yet one small gesture signals potential relief. As the waters start to rise, GOD, not Noah, gently extends His hand and secures the door of the ark. We sense that GOD wants to make sure that nothing happens to this precious cargo. From this small, simple yet majestic act we can conclude that GOD is securing Man's future!

As time passes and the flood comes and goes, GOD has time to reconsider His actions. When the ark hits land, GOD puts into motion some of Judaism's most beautiful and memorable concepts. GOD declares:

"And the waters will not become a flood to destroy all flesh again... And the rainbow will be in the cloud. And I will see it, to remember an eternal Covenant between GOD and every living being of all flesh, that is on the earth!" (Gen. 9:15-16). (We will explore the extraordinary importance of the word "Covenant" later on in this book.)

What I find remarkable about this is that GOD says that the rainbow is for Him to remember what He has done and, further, to remember that He is committing never to take such action again. It is GOD Himself that initiates a symbol for all of humankind to know. As the symbolic parent, He must make the first demonstrable move if the child is to take this relationship seriously.

Thus, the rainbow is there for all to see and, equally as important, for GOD to remember His pledge. How extraordinary is this?

Do you know anyone who would make such a public statement acknowledging his actions and openly sharing his commitment with the entire community? Do you know anyone who would specifically indicate the terms of that commitment by voluntarily issuing a public statement, on the record, from which to be forever judged? I personally do not know anyone who would have or could have gone to such lengths to set things right with his children!

Would you act as GOD did? Could you?

This early lesson is one that GOD establishes for Himself as well as Man! He, as well as we, should not forget that human life is precious.

The next time you see a rainbow, it will surely have new meaning!

CHAPTER 12
THE COVENANT WITH NOAH - THE DETAILS

Although GOD takes the first step with His public announcement, He recognizes that humanity surely will need additional, direct guidance.

So how does GOD proceed? He provides Mankind a structure of guidance in the form of a Covenant. We all have heard the term "Covenant" and have some general notion about what it means. However, the Covenant in the Book of Genesis has a more wondrous, far-reaching significance. This structure was not chosen at random, not simply "picked out of a hat." In approximately 2500 B.C.E. it was customary for the Hittites and other ancient near eastern societies to use the covenantal form of understanding to formalize acknowledgements between groups.[1]

As unlikely as it may seem, this formal structure actually assists in perpetuating Judaism throughout the ages. It keeps in place a process of reaffirmation of principles, resulting in Judaism's continued relevance and, most wonderfully, engenders a spirit of institutionalized hopefulness. The Covenant allows Man to continue his maturation while always being surrounded and encouraged by GOD'S inspiration.

Does this sound too good to be true? Can GOD's guidance found in the Covenant really provide all of this? Well, let's continue and find out.

We will begin by reviewing the details of this first Covenant and, in the next chapter, analyze the all-im-

---

1   Levine, *The Old Testament.*

portant effects of the covenantal structure itself.

"And God blessed Noah and his sons and said to them, 'be fruitful and multiply and fill the earth. And fear of you and dread of you will be on every living thing of the earth and on every bird of the skies, in every one that will creep on the earth and in all the fish of the sea. They are given into your hands. Every creeping animal that is alive will be yours for food: I've given every one to you like a plant of vegetation, except you shall not eat flesh in its life, its blood, and except I shall inquire for your blood for your lives. I shall inquire for it from the hand of every animal, and from the hand of a human. I shall inquire for a human's life from the hand of each man for his brother. One who sheds a human's blood: by human his blood will be shed, because He made the human in the image of God. And you, be fruitful and multiply, swarm in the earth and multiply in it. And God said to Noah and his sons with him saying, And I: hear I am establishing my Covenant with you and with your seed after you and with every living being that is with you, of the birds, of the domestic animals and of all of the wild animals of the earth with you, from all those coming out of the ark to every living thing of the earth. And I shall establish my Covenant with you. And all flesh will not be cut off again by the floodwaters, and there will not be a flood again to destroy the earth. And God said, this is the sign of the Covenant that I am giving between me and you and every living being, that is with you for eternal generations. I've put my rainbow in the clouds and it will become a Covenant sign between me and the earth" (Gen. 9:1-13).

In these 13 verses GOD delivers some of Judaism's most sacred and cherished fundamentals:

- Be fruitful and multiply! This one is easy. There

were only a few humans left, so Man had better get started.

- Mankind will have dominion over the animals and all else on the earth. This reestablishes that Man is made in GOD's image. As difficult a concept as this has been to fully define and appreciate, this makes it clear that Mankind has the responsibility and the burden to remember and exemplify godly intent. By securing his free will, Man chooses to eat the fruit of the Tree of Knowledge of Good and Bad and, since he does not subsequently use his newly acquired capabilities wisely, GOD now begins to outline a moral code that, even today, continues to support many fundamentals of western civilization.

- Having recently eliminated just about all of Mankind and immediately recognizing that He will never do that again, GOD provides a rainbow as a sign signaling His commitment to His side of the covenantal agreement, in which He emphasizes that innocent human life will be deemed precious!

- To ensure that Man will always be aware of the sanctity of life, GOD now states that the blood of an animal, the essence of its physical life, is never more to be eaten. This is the first of the kosher laws that remain today. Every time we sit down to eat and thus sustain our physical life, we are to be reminded of GOD's requirement of respect for the sanctity of life. In order to have this become part of our everyday existence, GOD forbids us from partaking of the blood of an animal since it is the blood that sustains the life of the creature.

This is especially poignant because many of the ancient near eastern cultures specifically included blood in their eating rituals, as it was seen as a step on a path toward their attainment of immortality. The Hebrew GOD was stating the exact opposite: since it is blood that sustains life, we must not take life for granted. Therefore, the removal of the life-giving blood reminds us, each time we sit to extend our own life as we eat our meal, that life is never to be abused. Life is to be ultimately respected. How wonderful it is to infuse the idea of the preciousness of life into the midst of cultures that originally built their drive for immortality upon the taking of another's life-giving blood.[2]

As I finished studying this portion, I became more aware of just how personal a journey writing this book had become. I realized that my three sons were actually ahead of me in some regards. Each of them, on their own, had already demonstrated belief in the sanctity of life by choosing to not eat the blood of an animal. By pushing their personal boundaries, they decided to demonstrate their understanding of the sanctity of life on their own by eating no meat at all; each had chosen to be a vegetarian.

This section of the Book of Genesis also held a surprise for me. As a determined GOD continues to teach respect for life, He states that anyone who commits premeditated murder should be put to death. I was not expecting to find the edict regarding murder included in the Covenant with Noah. I had always assumed that this came later, as part of the Ten Commandments, well

---

2   Telushkin, 103.

into the Book of Exodus. Additionally, in my supplemental readings I learned that premeditated murder is so egregious to the Jewish soul and so important in the Bible that it is the only statement of the Covenants found in each of the five books of the Torah: Gen. 9:6; Exod. 21:12; Lev. 24:17; Num. 35:16; and Deut. 19:11-13.

Rabbi Telushkin writes, "The Bible regards innocent life as being of infinite value. One who murders an innocent person has committed an infinite evil!"[3]

The directive against premeditated murder is all-inclusive:

"If anyone beats a slave and the slave dies that person is to be put to death" (Exod. 21:20).

Slaves are not chattel; murder of any person is not permitted! All innocent individuals are sacred. The ancient Hebrews were instrumental in ensuring that justice and fairness were the rule for all individuals—not just for the Hebrews, not just for the powerful, not just for the rich, but for everyone!

To emphasize the importance of recognizing the sanctity of life, the Bible goes even further:

"On the word of two witnesses or three witnesses shall the one who is to die be put to death. One shall not be put to death on the word of one witness" (Deut. 17: 6).

In order to pay the penalty of being put to death, someone who has been accused of murder must first be confronted by two eyewitnesses to the crime. Yes:

---

[3]   Telushkin, 407.

not one, but two. Murder is horrid enough, but convicting and then putting to death an innocent man is even worse. So the Bible makes it very difficult to conduct a perfunctory execution of anyone. It should be noted that it is extraordinarily difficult to have two eye witnesses to a murder come forward and individually state the same details. Therefore, in reality, few executions of accused murderers ever really took place.[4]

These were the initial concepts of our first Covenant, the Noahdic Covenant, upon which GOD started us on our journey toward godliness and justice. They are still today, thousands of years later, the backbone of our staunchly held beliefs regarding the sanctity of life.

Life is to be regarded as Man's ultimate privilege as well as his ultimate right established under GOD's direction. How marvelous indeed that this idea of the sanctity of life is the foundation of our moral code and the starting point for our search for godliness.

As a people, we should be proud to have been at the forefront of the establishment of this wonderful concept.

---

4   Telushkin, 408.

CHAPTER 13

THE COVENANT WITH ABRAHAM -
THE ALL IMPORTANT STRUCTURE

God could have simply said to His people, "Do these things because I say so!" But, Man had not behaved very well when just being "talked to" by God; therefore, a more tangible method of understanding was required. The covenantal structure served as a contract between parties of relatively different positions and responsibilities and clearly defined expectations on both sides.[1]

Because the peoples of the ancient near east lived, communicated and traded among all of the lands of the region,[2] even the ancient Hebrews were most likely well-acquainted with a covenant as an established form of agreement between parties and, thus, would have understood a covenant's significance and requirements. It was the perfect medium to use in order to delineate what one party expected of the other. When the peoples of this age would say that "they cut a covenant," what they were saying was "we just made a deal."[3]

The structure of the Noahdic Covenant was a rather simple one. The more influential party (God) says, "If you do this, I will provide that," to the lesser party (Man). This is called the Royal Grant Covenant Structure.[4]

As Mankind matured, this initial style of covenant was no longer sufficient to meaningfully encompass the more complex interactions between Man and God. Therefore, the much more detailed Suzerainty/Vassal (S/V) Covenant model, also called the Abrahamic Mod-

1   Levine, *The Old Testament.*
2   Ibid.
3   Ibid.
4   Ibid.

el, was established.[5] In this model, the Suzerain is God, and the Vassal in Man. (See Addendum II for a more detailed explanation.)

To me, one of the most significant aspects of this new model is that the Suzerain must stipulate His requirements and standards, while the Vassal may decide whether to follow these edicts. As a result, Man becomes a full participant in the process! Further, and importantly, as the agreeing party Man must periodically have the text of the Covenant, the Torah, read to each succeeding generation. Thus the people of each generation must decide for themselves whether to agree to the details of the Covenant in order to have the agreement continue.[6]

It is this recommitting that I find so vital to Judaism's past and so crucial and meaningful to its future, Why? The answer will become clear as we next examine the differences between this covenantal style of contractual understanding and the formal legal structure of laws typically governing a municipality, state or country:

1. As mandated by the S/V covenantal procedures, the Hebrews' descendents must periodically relearn the Torah and only upon rehearing, rereading and relearning can the next generation reconfirm the people's acceptance. Thus, each generation is to know the words of the Torah and decide whether to inculcate these lessons into everyday life. After its own study of Torah, each generation can recommit to live a moral

---

5   Ibid.
6   Ibid.

life and to strive toward the goal of living in GOD's image. It is through this process of recommitting that the Covenant continues!

2. If the Bible had established its edicts as parochial laws, these laws would have applied only to specific geographic areas, changing at every boundary line. The laws of one country do not apply to an individual leaving that country; he lives by the laws of his new home nation. The covenantal structure however, allows an individual to take his commitments with him wherever he goes; the Hebrew honors his commitments wherever life may lead him. The Jew and his Judaism are not restricted to any one place.[7]

3. In a national legal system, society's representatives establish the standards for everyone's behavior. [8]In a covenantal structure, each individual personally agrees to the standards by which he is to live—standards that commit Man to strive toward increasingly moral behavior. Therefore, it is the structure itself that dictates that Man may choose a lifetime of performing actions directed toward achieving the goal of heightened godliness. The Hebrew voluntarily agrees to the ground rules rather than having a state or country impose its will. By signing on the dotted line, Man chooses to live a life of increased righteousness.[9]

4. The individual has no direct role in his own government when living within the laws of a state or country. The statutes are passed and enacted by others, and

---

7   Ibid.
8   Ibid.
9   Ibid.

the citizen must obey them. In the S/V, or Abrahamic, covenantal structure, the individual is an active participant, which develops a living, breathing relationship between parties. The necessary partnership between God and Man also ultimately affects Man's relationship with Man as well.

The significance of this is that a Jew is not necessarily a Jew because he is told that he is a Jew, but because he has agreed to be one! He has heard the Torah, has rededicated himself to finding holiness and, in so doing, acknowledges that only then does God's Covenant have complete merit to him and for Him.

The process that requires each succeeding generation to recommit itself to and find merit in the Covenant is important for an additional reason. By definition this continual reevaluation proceeds within the context of the world in which each generation is living. The Jew studies and understands the Torah's lessons within the context of his current cultural, societal and religious environment. Therefore the words of the Torah that are reheard and restudied by each generation will always remain the same, but their manifestation may gradually evolve from era to era.

For example, while the Torah forbids us to work on the Sabbath, it does not define what work includes.[10] Each generation is to consider the significance of this for itself. It is through this process of restudy, reexamination and then recommitment that Judaism retains its vigor and relevance over time—for all time.

---

10   Matt, 94.

It now becomes clear that the Covenant provides the structure that keeps Judaism vibrant, alive and focused from generation to generation. Each of us is to relearn and recommit to Judaism's standards and, by reflecting upon these standards within the context of the world in which we find ourselves, we can continually help to guide the world toward godliness!

The Covenant contractually obligates each generation to look Judaism squarely in the eye, to decide whether to use the Bible to approach holiness and, if so, to pursue holiness through the means that God has laid out. Restudying God's roadmap keeps our people focused upon taking the actions necessary for Mankind to help achieve its greatest level of godliness.

Furthermore, the structure itself allows for the possibility that one generation may not necessarily completely agree with a former generation's set of priorities. This, however, does not indicate any lessening of, or wavering from, the goal of every generation's striving to live in GOD's image.

So, a GOD that we cannot completely understand has established a structure within which we may continually participate. God has provided guideposts that can lead us to greater respect for our fellow Man and that, we hope, results in an increased sense of the awesomeness of the essence of GOD. This process keeps us grounded in a relationship with God by focusing Man, in every age, to struggle to do what is effectively right and stay away from what is clearly wrong. It sets up the conditions that direct us to continually search for ways to truly live holy lives.

If Man is capable of knowing that we are all equally made of the same "stuff," then each of us might choose to slowly *let go* of our negative aspects and find ways *to let into* our consciousness the wonderful, positive aspects of the Torah's approach toward godliness. Thus, it is our *choice* to accept our role in the Covenant with GOD that can enhance our attainment of universal oneness, making holiness closer at hand for everyone.

Considering that the Jew always has the opportunity to choose a path that leads toward increased godliness, through so choosing he recommits to living in GOD's image and, in essence, to helping to make tomorrow better than today. Therefore, and most crucially, it is the very structure of the Covenant that places Judaism firmly upon a platform of institutionalized hope!

How, then, can atrocities such as the Holocaust or Darfur occur? It is not that GOD has thrust such evil upon Mankind but, rather, these are the tragic results of Mankind's failure to have sufficiently learned how to be truly committed to Man's side of the bargain.

Man has let Man down! It is Man who enacts these horrors upon Man—while God weeps!

Mankind does not always live up to his commitment to approach godliness, as Man has yet to truly understand that we are all of the same "seed." Yet it is always within Man's power to use the guideposts provided by God to help move all of Mankind toward peace.

The beauty of this Covenant is that its structure prevents Man from forgetting what the phrases "never again" and "in GOD's image" can truly mean to all of Mankind, today and in the future. When this opportunity is taken seriously, the Jew lives his life constantly knowing that he is capable of making a positive difference. Thus, he always has the potential of manifesting the precepts of institutionalized hope!

This GOD of the Hebrews is pretty clever.

This layered understanding of the meaning of the Covenant is what I call a "nexus point"—the point at which GOD's awesomeness intersects with humankind's limited capacity to relate to God. It is a point at which the unknowable and relatable converge. It allows Man to grasp God's road map while never deluding himself that he will ever know all of the complete essence of GOD.

Although the covenantal structure may create certain tensions between some of the understandings of one generation and some of those of the next,[11] it is the prescribed never-ending restudying and re-searching that is precisely what keeps Judaism focused upon achieving a path toward holiness, now and in the future. The brilliance of this construct makes Judaism as relevant today as at any time in our history. It allows Judaism to continue to be the world's moral lens for generations to come. We can create a beacon for the world by leaving our egos and prejudices behind and take steps to demonstrate that Man should respect Man on his way to godliness.

However, that's only if we choose to do so!

11   Matt, *God and the Big Bang.*

Therefore, a Jew is a Jew because he chooses to be a Jew as he formally accepts to search for something that he senses is larger than him. When Man chooses not to truly recommit to a path of godliness, Mankind pays the price, as Man ends up harming Man.

The satisfaction and hopefulness that I feel when I study this more complex covenantal structure of the Torah come from knowing that, at any time, anyone and everyone can decide to personally recommit to upholding the marvelous standards of the Bible and, as a consequence, the covenantal structure can be responsible for continually revitalizing Man's search for, and march toward, goodness. Judaism helps to look past today and focus on tomorrow! One person's very next words or actions can make a difference and help to guide the world toward godliness. As Daniel Matt writes, "The Holiest moment is now!"[12]

The Jewish spirit is continually refreshed by the winds of hope that are inherent in our committing ourselves to be part of God's road map. All of this comes from considering the significance of a single word in the Bible: "Covenant"!

"Institutionalized hope" —I like the way that sounds.

---

12   Matt, 100.

CHAPTER 14
"THE CHOSEN PEOPLE" - IT'S ALL IN THE CHOOSING

As we approach the episodes involving Abraham, it's important to keep in mind how important the Covenant is to the structure of Judaism. The Jews of each generation are to agree anew to the details of the Covenant, and it is the act of agreeing that is so vital and engenders such significance.

In essence, we choose to be Jewish. The Covenant's continued efficacy rests in our "choosing."

We seem to continually bump into the words "choosing" and "chosen." Every time I say or read these words a little bell goes off inside my head as I recall a phrase that has been one of the most misunderstood concepts of our entire history: "the Chosen People."

The conventional wisdom of many outside of the Jewish world, as well as of some within the Jewish world, assumes that the phrase "the Chosen People" implies that the Hebrew people believe that they are somehow superior to others.[1] This incorrect interpretation has unfortunately led people throughout history to ask, "Who are these people who believe that they are chosen by GOD to be superior to us? Who do they think they are? What makes them so special?" These misguided feelings have sometimes led to a sense of resentment toward our people and a pervasive, ingrained mistrust of our religion and culture.

When taken out of biblical context and read without study or thought, this phrase seems to imply that

---

1   Artson, 13.

the Hebrews believed that GOD would treat them with special favors and blessings, that they would have divine assistance and, thus, that life would be more carefree and engender a sense of superiority.

After my studying, I believe that it turns out to mean almost just the opposite!

What "the Chosen People" really indicates is that the Hebrews have a most solemn responsibility: to demonstrate to the world what it is like to live lives guided by justice and righteousness. We have been "chosen" not to take it easy or to have a "leg up," but to carry on life's most burdensome task: to exemplify, through our actions, what it means to live in GOD's image!

This is surely not an easy task. Since through his actions Man chose free will to acquire knowledge, God expects us to use this "gift" ethically and to uphold our part of the bargain. Therefore, "The Chosen People" appellation is not frivolous by any means. The Torah demonstrates how difficult it is for Mankind to live within God's standards.

Thus, just as GOD has chosen us, we must concomitantly choose to exemplify God's standards by the choices we make in every aspect of our daily lives. GOD has "chosen" to see whether we choose to live in a godlike manner by making choices that all of Mankind, both Jew and non-Jew, can emulate.

In his book *What the Jews Believe*, author David Ariel says, "The Jewish people have the duty to be a light of nations and bring God's teaching to the nations of the world."[2]

As we are finding out, being Jewish is not easy. Over and over, the Torah juxtaposes glimpses of GOD's image against Man's negative actions, demonstrating just how far Mankind is from pursuing life "in GOD's image." God helps by setting guideposts for us to follow, but it boils down to turning "the Chosen People" on its head: we should read this as "the *Choosing* People."[3] We must regard it not as GOD's having chosen us but, more appropriately, as if we have chosen to manifest godliness by our choosing to live by standards that are godly inspired. In essence, we are choosing Him as well, as there is no challenge greater, nor burden heavier, for Mankind than being "chosen" in this manner.

Being "the Chosen People" carries with it a set of both real and implied responsibilities bestowed upon us to honorably carry out; it is not an honor devoid of God's required actions. Let us choose to be "the Chosen People," a people who choose to lead in the pursuit of justice, mercy, righteousness and holiness.

It is a choice worth making!

---

ism (New York: Schocken Books Inc. and Random House Inc., 1995), 117.
3   Ariel, 120.

CHAPTER 15
WHAT'S IN A NAME? I WOULD SAY QUITE A LOT

Man has a tendency to label and name things.[1] By doing so, we create a reference point from which to begin to relate to what we find around us. A name places things into a context of general recognition.

When I say "table," everyone knows, in general, what I mean. However, the name "table" does not tell us anything about the object except that it has some kind of a top and that this top is held up by something underneath.

Let's take the word "tree."[2] Everyone knows that trees have leaves and trunks. But, what really is that tree? How many thousands of different kinds of trees are there? How do their root systems vary? How many species of insects and animals live within one set of trees and might not live in others? Do all trees make sap? Is all sap edible? How many different designs of leaves are there? Do all trees make nuts? Which trees die in the winter? How many different types of barks are there?

The truth is that there are so many variables among different types of trees that the name is just about meaningless beyond its broad description. The label "tree" is really meaningful only when considering the most common characteristics of all trees.

Now think about standing by what we label a "river." You see water flowing within the river's banks. Take a 10-minute walk in the nearby woods. When you

---

1   Matt, *God and the Big Bang.*
2   Matt, 38.

return to the river, it seems exactly the same as the river that you left 10 minutes earlier. Yet, it is surely not. Thousands or millions of gallons of new water have replaced the water that you first saw. The current has replaced old tree branches with new ones. New fish now swim where other fish had been but have moved on. The river that you now see is surely different from the one that you first experienced.

Whether we're talking about trees in general or a particular river on a particular day, our names and labels take into account a broad and fluid set of characteristics.

Now let's talk about Man, specifically about someone named Saul. Initially, we should understand that Saul's skin replaces itself every two years.[3] Every single cell of the 10 trillion in his body is replaced every 10 years.[4] Therefore in the most basic sense, Saul is truly not the same man he was 10 years ago.

As Saul ages, his environment and hormones change so the reactions that Saul had in the past to a particular set of circumstances are not necessarily the same ones he may have today. This "allows for no permanent abiding self."[5] Additionally, Saul has the ability to present the image to the world that he feels will be judged most favorably, even if it does not necessarily reflect his true inner character.

As people meet Saul at varying times of a day or throughout different periods of his life, each of these individuals will encounter a different Saul. The extenuat-

3   Bryson, *A Short History of Nearly Everything.*
4   Ibid.
5   Matt, 63.

ing conditions of these meetings may have been so different that some actually may come away with opinions of Saul completely opposite to those held by others.

The result is that our need to place an arbitrary label on things and people instantly limits the true and complete nature of what we are naming! A name helps us to communicate with each other, yet this communication is not very consistent, accurate nor complete. You can see how easily miscommunication and misconceptions can occur if someone thinks: "The Saul that was described to me was the exact opposite of the Saul that I know."

Furthermore, even when we try to communicate with increased detail, how much does anyone truly know about any particular subject? What percentage of what we once knew of a subject have we forgotten? How much of what we are exposed to just passes us by?

"The human brain assimilates just one-trillionth of the information that reaches the eye."[6] It is not an exaggeration to say, "We have forgotten more than we currently know." The human mind sifts through these trillions of pieces of incoming information and "decides" to store what it believes should be the main focus. Therefore, the vast majority of what we could know simply washes through us.

These issues are compounded when we try to name GOD.

Labeling GOD would immediately and incalculably place boundaries around GOD. Because of Man's lim-

---

6   Matt, 61.

ited capabilities, any name or label that we assign to GOD would be an attempt to delineate what is in fact a divine mystery. In other words, if Man can conceive it, then it really cannot be all of GOD. As we have seen in previous chapters, Man is capable of relating to only a portion of GOD; therefore trying to define or name GOD would only diminish His completeness.[7] The attaching of any prosaic name, label or definition by Man to GOD does an untold amount of damage to the totality of GOD.

How does the Bible deal effectively with this problem? In a most wonderful way!

The Bible uses a tetragrammaton, a combination of four simple Hebrew consonants, to indicate GOD's presence. These four letters are "YHWH"; they are accompanied by no vowels. Try to pronounce how you suspect that these four letters, with no vowels, may sound.

What you end up with is a breath, a sigh, a whisper. The Torah makes GOD's "name" impossible to verbalize! Since it is unspeakable, it is not a label or name at all. GOD becomes knowable and hidden at the same time—recognizable, but unexplainable.[8]

The simple four letters "YHWH" form the root of the Hebrew infinitive "to be" as well as many of the conjugations of the verb.[9] The present tense forms are "am," "is" and "are"; the past tense forms are "was" and "were"; and the future tense form is "will be." In the first person, YHWH can mean "I am what I am,"

---

7   Matt, 101.

8   Artson, 20.

9   Artson, 20; Friedman, 178.

"I was what I was" or "I will be what I will be." In the second person, it can be read as "You are, you were, you will be." In the third person, it becomes "He is, He was, He will be." This "name" places GOD everywhere and in everything, unbounded by time place or person. Think about it: GOD has been, is and will be everywhere and everything.[10]

Thus, in its purest form, GOD is no specific thing. This makes sense because, if GOD were identifiable in the way that we know what we mean by "table," "tree," "river" and our friend "Saul," then GOD would be limited and diminished by this labeling or naming.

Therefore, GOD is actually...No Thing![11]

GOD is NoThing! God is Nothing.

GOD's grand essence is found in His Nothingness![12] Our GOD of everything is found in this No Thingness! He is everything and everywhere and, therefore, not defined by any specific thing; He is No Thing!

As Man grasps the importance of GOD's Nothingness, he has the opportunity to also minimize his own sense of self-importance and to become less anthropocentric. The less Man thinks of himself as a special specific entity, separated from the rest of the universe, and the more Man understands that we are all made of the same stardust, the closer we all will be able to understanding our potential oneness with the GOD of everything. We can then become more a part of every-

10   Artson, 20.
11   Matt, 40.
12   Matt, *God and the Big Bang.*

thing with an increasing understanding and awareness of our oneness with the universe and with GOD.

The more a man perceives of his own No Thingness, the more like GOD he can become.[13]

Not through a scream, a declaration, a miracle, histrionics, a demand, a threat or force does the Bible acknowledge GOD's existence. The Bible makes this acknowledgment through the exquisite subtlety of a breath, a whisper, an audible sigh.[14] It is as a "still small voice." GOD's non-name draws us directly toward the understanding that the GOD of everything cannot be tied to the limits of time and place.

Now, let's take a second to go back to the cosmological Singularity and the moment of the unknown immediately prior to the Big Bang. This moment signified a state of nothingness. From this nothingness, the Big Bang occurred, resulting in everything we have in our universe—even us!

For approximately 13.7 billion years now, the universe has been expanding. Within this expansion an almost infinite number of occurrences had to have happened for Man to end up as we know him today. Everything we see and know is directly traceable back to the Big Bang and, more interestingly, all of the way back to the state of nothingness prior to its initiation. Whatever it was that put the Big Bang into motion is also responsible for our reality today—a reality clearly embodied in the oneness of everything that began out of a state of nothingness!

---

13   Ibid.
14   Artson, 20.

This provides an increased understanding that science and the study of religion are not mutually exclusive; one neither excludes nor precludes the existence of the other. In fact, in this context, science and religion are complementary.

The Book of Genesis captures this total essence of the GOD of everything physical and spiritual by using a non-name that underscores His limitlessness, by not allowing GOD to be linked to any specific thing. Thus, I view the GOD of Judaism as possible, ubiquitous and awe-inspiring, and I am aware that I am part of the oneness of the universe that came from both cosmic nothingness and spiritual No-Thingness.

I find this wondrous and beyond complete comprehension.

Through the use of four simple letters, the Bible helps human beings to understand the existence of everything and our ultimate link to godly No-Thingness. The more that I understand the significance of GOD's No-Thingness, the more I cherish how the Bible has so marvelously handled the issue of GOD's name. Through the use of four simple Hebrew consonants and the omission of vowels, the Bible presents a word that cannot be spoken, yet represents the presence of the GOD of everything—of yesterday, today and tomorrow!

So what's in this name? I would say quite a lot!

CHAPTER 16

ABRAHAM- THE FIRST JEW

We are ready to take our next step as Genesis moves from GOD's involvement with the universe, the earth and Mankind as a whole to more directly addressing the individual. For this important next phase, whom does God choose? Abraham. The Bible does not specify why God selects Abraham. This is an example of an instance that requires the reader to engage with the text and come to his own conclusion.

"And YHWH said to Abram, 'Go from your land and from your birthplace and from you father's house to the land that I will show you. And I will make you into a great nation, and I will bless you and make your name great. And be a blessing'" (Gen. 12:1-2).

Initially, the relationship between God and Abraham resembles the relationship between GOD and Noah. As with Noah, Abraham is summoned by GOD. Abraham is told to 1) leave his birthplace, 2) leave his home, 3) leave his father and 4) pick up and go to a strange land to which God will lead him and which turns out to be Canaan. (There is some debate over Abraham's original birthplace; it is either The Great City of Ur in southern Iraq or the city-state Ur of Chaldees, formally know as Urfa, in the southern part of Turkey).[1] In Abraham's case, the reason for the summons is that God wants to tell Abraham that He will make Abraham's descendants a great nation. Additionally, Canaan "represents a shift in consciousness, a place that can nurture deeper spiritual insight by virtue of the events and institutions that will emerge from its soil."[2]

---

1   Rendsburg, *The Book of Genesis.*
2   Artson, 16.

So Abraham listened and went to Canaan!

We see here a human being who is obeying God's directions. Even though it means leaving everything that is dear to him, Abraham simply makes the necessary arrangements, picks up everything and, like Noah, moves without a sound. There is no questioning or any input as to why or how at this point.

However, get ready—the relationship between God and Man is about to change forever!

Actually, at this point we are at the edge of change where all relationships—GOD to Man, Man to God and Man to Man—are about to become deeper, richer and more complete. These relationships will be tested and will become increasingly complex as the parties throughout the Book of Genesis bump up against the reality of human frailties. For the remainder of the Book of Genesis, all of the participants will struggle to find ways to function while confronting the consequences of their actions.

Once in Canaan, Abraham finds that the land is experiencing a famine, so again he packs up his family and belongings, and this time he decides to head to Egypt, where there is sufficient food for him and his minions. With him are the members of his extended family, including his nephew Lot.

The group makes its way to Egypt. After some time, Abraham and his family find themselves healthy, wealthy and able, and they decide that it is time to return to Canaan. Once back in Canaan, Lot announces that the land does not *"suffice them to live together"* (Gen. 13:6).

It seems as if it was fine for his nephew to live with Abraham and his family when everyone was poor and hungry. However, now that Lot's family is financially secure and his household has expanded, he seems to say that he no longer needs Abraham.

Upon hearing Lot's request, Abraham tells his nephew to choose where he wishes to move, point in that direction and go. So off Lot goes—ironically, it will later be revealed, to the area that includes Sodom and Gomorrah.

For many years prior to this moment, God has repeatedly told Abraham:

"I will make your seed like the dust of the earth, so that if a man could count the dust of the earth then your seed also could be counted" (Gen. 13:16). (Again, the word "dust" is used. The significance of dust is universal.)

However, after these years of being assured that he and his wife Sarah would have an heir who would follow in his father's footsteps and lead the Hebrew people, and years of having his hopes routinely dashed, Abraham decides that it is time to confront God. The Patriarch's continued frustration and disappointment reach a climax, and he simply snaps. He has left his father and moved to a strange land that was in the midst of a famine, forcing him to move again. He has worked to support his extended family in a strange land only to be informed by his nephew that he is not needed anymore! And all the while, for these many years, he has endured the sadness of the unfulfilled words of God: Do not worry, as your children will populate the earth and carry my message to the four corners of the globe. Yet there has been no child! As Abraham and Sarah find themselves aging, the frustration just becomes too much for Abraham to bear silently.

So in Genesis 15:2, Abraham cannot contain himself any longer. This changes everything!

"And Abram said, *'my Lord YHWH, what would you give me when I go childless?'*"

And in Genesis 15:3: *"Here you have not given me a seed."* As Abraham asks God why He has left him childless, you could cut the tension with a knife. Abraham is confronting God!

I am once again struck with how personal this journey is for me. Since Susan and I had difficulty having children, with this story I find myself relating completely to the biblical couple's anxiety and frustration. I know this tension and anguish and how intense it can become.

For Abraham, there is no turning back the clock now! For the first time, a human has dared to question God! The relationship has now matured; the ability of the Jewish people to enter into a dialogue with God is now established. Man has begun to fulfill his covenantal responsibility. In his search for guidance, he becomes an active participant. The covenantal requirement that each generation must reexamine and question the Covenant now takes its first cognizant step. Man has asked God the first of what will be a never-ending series of questions. This question is asked in sincerity and within the context of the circumstances in which Abraham finds himself.

After this first question, contending with GOD becomes part of the Jewish tradition.[3]

The capability of Mankind to both question God and search for answers was made possible by our gaining free will, which was acquired by our eating from the Tree of Good and Bad and extended by the covenantal responsibility to study, question and re-commit to the understanding of God. It has taken this long for this capability to be actuated.

And by whom? By none other than Abraham. The first Patriarch. The first Jew! Who else?

---

3   Telushkin, 32-33.

CHAPTER 17

THE VIRTUE OF ABRAHAM - TRUST

Now that Man has demonstrated his capacity to question God, let's consider Abraham's specific questioning. As Abraham ages, he is no longer able to contain the frustration of not having been provided with the promised heir, someone to carry on his work. We find an Abraham who simply cannot cope with this frustration any longer and, after years of absorbing the disappointment of GOD's unfulfilled assurances that a son will be provided, decides to challenge God directly.

How does God react to being challenged in such a manner? What happens next is very important!

As He has demonstrated with Noah, GOD is capable of reacting harshly toward humankind. Although in that episode GOD reacts because Man has been misbehaving and not, as now, simply challenging God, a precedent has been set. But, like Man, God has matured within this relationship with humanity. With Abraham, God responds calmly.

"And he brought him outside and said, 'look at the skies and count the stars, if you will be able to count them. That is how many your seed will be.'" (Gen. 15:5).

No anger, no threat, no "how dare you ask me such a question!"—just a simple restatement of the same position he has explained to Abraham many times. Calm reassurance, no histrionics.

So the ball is back in Abraham's court. We now

come upon what will be the hallmark of Abraham's virtue: trust!

"And he [Abraham] trusted in YHWH" (Gen. 15:6).

Again Genesis uses simple words packed with unimaginable power. But why does Abraham trust God? Upon what is this trust based? How can Abraham trust God after the promise has gone unfulfilled all these years?

It is precisely because God has calmly allowed the confrontation to occur that trust can develop. God does not become enraged or demand that Abraham recoil in fear for having asked a question. The maturation of Man's relationship with the relatable part of God continues.

God tells Abraham that a seemingly impossible set of occurrences will take place "through" him. Abraham has challenged God on this issue, and God's reaction is a simple reiteration of this seemingly impossible feat. Abraham does not receive a haranguing retort from God. No threats, no fire and brimstone. However, since God has not provided a child for Abraham up to this point, which would have been proof positive of God's promise, Abraham has a choice to make. He may be saying to himself, "God has responded to my questions. Now how do I proceed?"

Abraham chooses to trust in God![1]

"And he trusted in YHWH and He considered it for him as virtue" (Gen. 15:6).

---

[1]   Artson, 18.

Abraham returns God's trust. Abraham must realize that God has trusted in him from the beginning, choosing him to be the first Jew, the first Patriarch. And, as the first Patriarch, his heir would assuredly be there to continue to extend GOD's Covenant.

Abraham decides to say nothing more. With God's reaffirmation of his promise, the discussion ends. Abraham will wait to see how this will play out.

Yet Abraham is not wholly confident in the outcome. Implied in the word "trust" is the acknowledgment that some doubt still exists,[2] that some uncertainty may still surround the issue. Abraham chooses to look beyond these doubts in order to move forward.

Trust is not simple acquiescence to an idea or the stifling of doubt. Confronting doubts, realizing that doubts linger and then acknowledging these doubts enables us to move on and not get stuck. It is through the process of integrating inward struggle with outward experience that helps Man to build layers of trust that eventually establish the foundation upon which trust can possibly grow into faith. Jewish faith is not blind. We struggle to challenge our doubts and find as many answers to our questions as possible.[3] Because of Man's limited capacities, we do not have indisputable answers to some questions and, thus, some doubt does exist.

In our everyday life, we have come to accept some level of doubt. Cosmological examples, too, illustrate the way we trust that certain things are "true" even though scientists cannot determine the *why's*:

2   Ibid.
3   Ibid.

• We cannot see super massive Black Holes, and we do not know why they first condensed, but science tells us that they do, in fact, exist. A super massive Black Hole allows no light whatsoever to leave its grasp. A telescope reveals a complete absence of anything viewable to the eye; it appears as if nothing is taking up this space. However, physics indicates to us that it is exactly the opposite: the Black Holes have a mass that is so dense that not even light is able to escape its gravitational pull. Physicists are positive of the existence of the Black Holes, because they can measure the effects that they have on the "objects of space" that surrounds them, not because anyone will ever directly encounter one in a convincing way. We simply trust that they are there.[4]

• By studying the results of the Doppler effect (wave shifts), scientists have concluded that the galaxies are racing away from each other and are doing so at ever-increasing speeds, yet they do not know why this is occurring. One widely held hypothesis is that there must exist a repelling force called Dark Energy emanating from somewhere, even though we have no idea from where.[5]

• Current physics dictates that the outermost stars of most speeding and rotating galaxies should be leaving the gravitational pull of that galaxy and shooting out into the heavens. However, observations show that these stars are staying put in their orbits around the center of their galaxy. To reconcile this, most physicists hypothesize—and accept—the existence of Dark Mat-

4   Tyson and Goldsmith, *Origins: Fourteen Billion Years of Cosmic Evolution.*
5   Bryson, *A Short History of Nearly Everything.*

ter, which would account for the increased amount of gravity required to hold these stars in place in their orbits. To date, however, physicists have not found a sufficient amount of mass that would produce the gravity required to hold these stars in their orbits. They accept that it is there; otherwise, the stars would be sailing out into the dark night of space. But no one has been able to spot it.[6]

It is similar with Judaism. Man can know only what he is capable of knowing. Acknowledging that we cannot know all of GOD affords us the opportunity to garner increasing levels of trust, which helps us to strive toward higher levels of faith.

Similar to the way scientists develop and accept hypotheses to explain their observations, as Man absorbs the insights and experiences of biblical Judaism, we become more comfortable with our rightful place within the relationships of GOD to Man, Man to God and Man to Man. This makes it easier for us to trust in these relationships; it moves us toward the possibility of experiencing increased faith in a GOD that we ultimately cannot completely know.

There is a great deal of everyday life that we simply trust will happen. When we walk into our homes and flip the light switch, we trust that the light bulb will turn on. When we look out of our front door and see dark clouds, we trust that the rain will fall.

So as we juxtapose our trust and the knowledge that we accumulate from life's experiences with the ac-

6   *Science Today* (November 15, 2002).

knowledgement that we do not always know the *why's*, we are able to begin to pursue a possible path toward faith. As we become increasingly aware that we do understand and experience many of the *how's* and *what's* of the universe, the leap from trust to potentially increased levels of faith can become less intimidating.

In this light, Abraham views God's reaction to his challenge as supportive and reassuring. He has learned something important about God and now sees no reason not to trust God's words. Most important, Abraham begins to demonstrate to us that with a deepened trust, faith may become easier to attain. The supportive manner in which God wonderfully reacts to Abraham's emotionally charged questions enables Abraham to increasingly trust in God, which in turn assists Abraham's movement toward faith in God.

This first patriarch was pretty sophisticated.

CHAPTER 18

We all know the children's story version of the wicked Sodom and Gomorrah, but what I did not realize was that the story actually begins when Abraham and Lot part ways.

"The land did not suffice them to live together, because their property was great. And they were not able to live together." (Gen. 13:6). So Abraham and Lot must now go their separate ways: that he is no longer able to live with Abraham and his people

Ironically, the area to which Lot chooses to take his family, the plane of the Jordan, includes the cities of Sodom and Gomorrah.

When God tells Abraham of His decision to destroy the cities of Sodom and Gomorrah because of the wickedness emanating from the cities, Abraham is aware that Lot and his family are living somewhere in that area. With this knowledge, Abraham enters into a debate with God in Genesis 18:26-33, which becomes the longest dialogue between parties in the entire Bible. Abraham challenges God not to destroy the cities if God can find at first 50, then 40, 30, 20 and, finally, only 10 virtuous inhabitants. What is significant here to me is that these inhabitants do not have to be Hebrews but can be non-Hebrews as well; thus, Abraham expresses concern for the lives of all men of the region no matter what their religious or cultural background.

Since eventually most of the people of Sodom and Gomorrah show themselves to be evil—some even surround Lot's home and threaten the angels of God that are visiting Lot—God ends up leveling the towns. The overarching point is, however, that Abraham's supplication to God to spare the human lives within the cities indicates that Man, through Abraham, may be getting his "ethical sea legs." Through Abraham, humankind is beginning to display an awareness of the "specialness" of human life and this awareness expands, nurtures, broadens and recasts itself within Man's relationship with God. Man has demonstrated that it is imperative to debate with God if the issue is as important as the sanctity of human life.

Further, even though Lot lacks appreciation for Abraham's efforts to help him provide for his family in Egypt, the Patriarch rises above this pettiness. Abraham knows that by asking God to save Sodom and Gomorrah, he also is assisting in saving Lot and his family—showing that sanctity of life trumps pettiness between individuals.

Most important, when Abraham decides to engage God in this discussion and challenge God to explain His actions, God deems this request worthy of acknowledgment and enters into the longest dialogue He will have with Man.

It is wonderful to watch as Man and God learn to relate to each other. We witness a significant moral maturation of Abraham. He has grown from having concerns that relate merely to his own personal situation, his childlessness, to arguing for the salvation of others and giving practical application to the concept of sanctity of life.

Abraham matures.

God listens.

The dialogue ensues.

The relationship moves forward!

CHAPTER 19

"THE HAGAR EFFECT" - DON'T LOOK NOW

As Abraham's wife Sarah grows certain that she will remain childless since she and Abraham are now in advanced age, her sense of inadequacy becomes all-consuming. She feels sorry for her husband Abraham and, keeping with the customs of this era,[1] urges him to "take" her Egyptian concubine Hagar and have a child with her.

A son, Ishmael, is born from this union. Ishmael is a blessed son of God and will play a truly significant role in Genesis. God speaks of Ishmael's future in Genesis 16:10:

"I will multiply your seed and it won't be countable, because of its great numbers." Although Ishmael is a blessed son of God, it turns out that he is not *the* blessed son of God, as that distinction is to be bestowed upon Isaac. God finally fulfills his promise to provide the aging Abraham and Sarah with a child, and their son Isaac becomes the second Hebrew Patriarch.

As Isaac grows older, it becomes increasingly difficult for Sarah to see Isaac and Ishmael together in the family, until one day Sarah decides that this situation cannot continue any longer. She even falsely accuses Abraham of having been responsible for the problem in the first place. Sarah demands that Abraham remove Hagar and Ishmael from the family quarters and send them out into the desert.

A heart-wrenching scene unfolds in the desert as Hagar cannot bear to watch her son Ishmael suffering

---

1   Rendsburg, *The Book of Genesis.*

from lack of water. In order to avoid this pain, she turns away, weeping in distress, and starts to move away from her son. God hears her sobbing and, after she tells God that the boy will die without water:

"God opened her eyes and she saw a water well. And she went and filled the bottle with water and had the boy drink" (Gen. 21:19).

Look closely at these words. They do not say that God makes a new water well! It says that God opens Hagar's eyes, and she then sees the well. The well is there all along; Hagar's anguish blinds her from seeing it.[2] Only after she is calmed by God's words can she collect herself sufficiently to recognize that the well has been there all along. Until that moment she is too overwhelmed by the anxiety and stress of the situation to be able to recognize what is right in front of her!

We have all experienced the "Hagar Effect"! From time to time, we become so myopic regarding our daily responsibilities and activities of living—filled with the stresses of expectations, anticipations, hesitations, cultural conventions, increased compensations, and general worries—that we can be blind to the wonders that surround us each second of every day. We become unable to recognize that which is truly important in our own existence.

Studying, learning and comprehending the teachings of the Torah can help to minimize the impact of the Hagar Effect while maximizing the appreciation, joy and awe that exist within our own lives and that,

---

2   Artson, 25.

in turn, allow us to experience the wonders that are always before us. Overcoming the Hagar Effect allows us to cherish each moment of our unique lives and to become aware of each miraculous day!

We should take a collective breath...relax...re-focus...on the life-sustaining wonders. We simply have to relearn how to "see" them.

The Torah shines a light on them.

Have you ever been blinded by the Hagar Effect?

CHAPTER 20
THE POTENTIAL SACRIFICE OF ISAAC -THE AKEDA

"Take your son, your only one, whom you love, Isaac, and go to the land of Moriah and make him a burnt offering there on one of the mountains that I'll say to you" (Gen. 22:2).

For as long as I can remember, I've been under the impression that the episode of the Akeda, in which God instructs Abraham to sacrifice his son Isaac, is viewed as the ultimate test of Man's faith in God. However, after spending a great deal of time on this subject, I discovered that the Akeda episode is yet another multi-level event, full of meaning and open to various interpretations. I've experienced several such revelations during my journey, but this episode is of particular significance to me.

Let's start at the beginning.

"And it was after these things. And God tested Abraham, and he said Abraham and Abraham said, I am here" (Gen. 22:1).

By using the word "tested" in the very first sentence of this important episode, the Torah tips off the reader to not necessarily take what comes next at face value.[1] We do not know at this point what the test is or what the consequences may be if Abraham fails the test. However, since we know it is a test, we are able to read the remainder of the episode within this context.

God immediately proceeds to drop the bombshell, and the test is defined: Abraham is instructed to take his son Isaac to Mount Moriah and sacrifice him!

---

1   Rendsburg, *The Book of Genesis.*

The reader is taken aback by these words. How can God make this request? But again, we are experiencing the give-and-take between God and Man, although this command to a father to sacrifice his son ratchets up the intensity of the dialogue between Abraham and his Maker.

How can God direct Abraham to sacrifice the very same son who represents the fulfillment of the promise God made to Sarah and Abraham? How will Abraham react in this situation?

This is the same Abraham who questioned God about the birth of his son Isaac and debated God about the dismantling of Sodom and Gomorrah. Abraham was vociferous toward God regarding the birth of Isaac and almost over-demanding regarding the people of Sodom and Gomorrah. So we might expect a statement of indignation by Abraham. But this also is the same Abraham who decided to trust in God by going to a strange land and by accepting the promise that an heir would eventually be born to him.

How will Abraham react when the subject is the sacrificing of his own son—by his own hand? What does Abraham say?

Stunningly...Abraham says nothing. He reacts with complete silence, utter quiet! Just like Noah!

Throughout my Jewish education, I was taught that this silence is a simple, yet ultimate, expression of obedience to and acceptance of God. If Abraham is able to sacrifice the child that he so cherished, Abraham is displaying complete devotion to God's Word.

However, I would like to modify this assumption. My interpretation now is that Abraham does not protest simply out of obedience to God. Although that may be part of his reasoning, I believe that Abraham also senses and "trusts" that this is in fact just a test. Abraham and Sarah go through years of severe longing and disappointment regarding the birth of a son and are initially given Ishmael. In order to please his wife, Abraham is forced to send this son out into the desert. Additionally, he watches Sarah succumb to the great sorrow of not having her own son as promised by God. After the couple endures depression and devastating agony, Isaac is miraculously born. God's promise to them has indeed been fulfilled.

Therefore, I believe Abraham knows in his heart that, after everything he and Sarah have endured, God will not take Isaac's life and therefore this must be a test—a test that Abraham trusts will end acceptably.

Abraham might reason, "There is no need to confront God on this issue. Without Isaac there will be no future Hebrew generations! Since God has told me that my future generations will be as numerous as the particles of dust, this must be a test. So why protest? Hold tightly onto trust or faith, and proceed!"

While the reader knows from the first sentence of this episode that this is a test, Abraham has to reason it out for himself.

So, what is the effect of Abraham's decision to continue on with no protest? Abraham has effectively flipped this episode on its head and turned it inside out. He has made the Akeda a test of GOD as much as a test of Man!

This episode implicitly poses questions: Can God be trusted to follow through with the promises that He made to Abraham, Sarah and the Hebrew people? Will God back out of his Covenant with them that the future will pass to Isaac? Is it God's intention to show everyone that He was truly in charge?

Still another level of trust is displayed in this episode: the trust of a son in his father. Isaac shows no outward signs of fear, nor does he question the actions or words of his dad. Isaac simply listens to his father.

This very dramatic episode puts in a nutshell the fundamental precepts that exemplify the nature of the entire Jewish religion: a nexus of trust, sanctity of life and human purpose embodied in the relationships of GOD to Man, Man to God and Man to humanity! Everyone is being tested, not just Abraham!

Other clues, too, indicate that all will be right with the world—clues that help the reader to know that Abraham suspects that this is indeed a test.

"And Abraham said to his servants, 'sit there with the ass and I and the boy will go over there and will bow, and we will come back to you'" (Gen. 22:5).

Even before Abraham and Isaac leave to go up the mountain, we see that Abraham instructs the young men accompanying them to sit and wait for their return. He tells them that "we" will be returning, not "I will come back"!

Furthermore, there is absolutely no mention of Sarah or of Abraham's telling Sarah of this potential tragedy. Some interpretations attribute this lack of communication between Abraham and Sarah to an implied faltering relationship. I view this differently. I believe that Abraham, knowing that this is a test, cannot and will not cause his wife Sarah any further heartache by even mentioning the possibility of Isaac's death. How can he put her through such torment after what she already has endured? If this indeed turns out to be a test, why would he cause Sarah any further stress?

As this chapter of Genesis comes to a close, it is important to note that this episode of the Bible involves participants with personal histories that carry some level of experience both in the world and with each other. Since they know something of each other, each has a basis upon which to place trust in the others. Each can and does build upon the trust of previous experiences.

Additionally, are tensions, apprehensions and fears being felt within each of our participants up on that Mount? Are our participants possibly feeling that their trust has been misplaced? Are they anxious that something may go wrong? I suspect that the answer to all three questions, and with respect to all who are involved, is "yes."[2]

Why, then, throughout the entire episode, is there not one word regarding any level of anxiety or fear on any participant's part? Because the reader knows that this is a test from the first sentence of this chapter. I find this to be one of the most interesting, important and

2   Ibid.

extraordinary dimensions of the story. Again, the Torah marvelously involves us and, since each of us is aware of the anxiety that is felt during a test, the words of the Torah tap into these human feelings. The Torah assumes that the reader is engaged with the text and will naturally anticipate how much questioning must have been going on inside the souls of each participant.

The stage is now set. Abraham, Isaac and God have extended their trust to each other and, because of the seriousness of the event, each participant's trust has been stretched to unusual limits. Even though we believe that this is a test, the internal turmoil of each of our participants must surely have been significant since the subject in question is the potential sacrifice of a precious child at the hands of a loving father.

Wonderfully, the sacrifice is avoided. Isaac is spared, and Abraham is relieved of having to commit such an act. God does not go back on his word! Isaac will indeed lead the Hebrew nation. Each participant has acted on the highest moral level. Despite tensions and doubts, Abraham and Isaac have trusted in each other and in GOD.

Any lingering doubts are subsumed into an increased level of trust; as we have learned, each positive trust experience makes it possible to move toward a higher level of increased faith.

Trust is hard. Faith is harder!

Being spiritually aware takes work but, if approached properly, there is nothing more rewarding.

We continue to see that Man is capable of finding that part of God that is relatable to him and therefore, piece by piece, he is able to increasingly sense the aspects of the essence of GOD that are ultimately unknowable to him.

Man cannot be GOD, but he can act in a godlike way. Eating of that fruit of the Tree of Good and Bad has given Man the ability to choose to search for this path. It is a difficult task, but it is absolutely worth the effort.

The Akeda episode highlights the importance of Man trusting GOD, God trusting Man and, equally important, Man trusting Man. The testing is for all parties!

Therefore, the ultimate questions for Mankind are: Do we each "test" ourselves by striving to live morally each day, irrespective of the circumstances we find ourselves in? Does this testing help us to garner sufficiently increasing levels of trust in order to perhaps, one day, acquire deeper levels of faith?

This is truly a test worth taking!

CHAPTER 21

THE CAVE OF MACHPELAH - THE HEBREWS HAVE ARRIVED

Sarah lives a long life and dies at the age of 127 years. As a relatively recent inhabitant of the land of Canaan, Abraham has no official place to bury his wife. So he seeks out a Hittite landowner and enters into negotiations with him in order to buy some land of his own.

This formal practice is very much like our own today, with the parties discussing the merits of the land before the owner quotes a price. Normally, the purchaser then enters into a series of negotiations, obviously hoping to purchase this land at a lower price.

However, when the Hittite offers Abraham the land at a relatively high price of 400 shekels, instead of negotiating a lower price, Abraham uses some of the fortune that he was able to accumulate in the land of Egypt and proceeds to pay top dollar for this land right then and there, with no negotiating. As was the custom in the ancient near east, Abraham has the option to buy only the physical land, only the trees on the land or both. Abraham purchases both the land and the trees and owns it all.[1]

What is important about this transaction?

The significance is that forevermore, there can be no doubt that the Hebrew people have formal title to land and the trees in the "land of Canaan." The Hebrews have officially arrived!

---

1   Ibid.

CHAPTER 22
THE BIBLE'S FIRST PRAYER - IT'S NOT WHAT YOU THINK

As Abraham grows older and sees that his son Isaac has no wife, he sends a devoted employee to the City of Aram in order to find a suitable wife for his son. The employee is possibly a non-Hebrew,[1] who most scholars believe is probably Eliezer.

Abraham does not give Eliezer any specific details or characteristics of the kind of woman that he would suggest he bring back for Isaac. We can assume that, since Eliezer has been in the employ of Abraham for so many years, Abraham must feel that he will be able to choose someone who will be acceptable for Isaac, the next leader of the Hebrew people.

Although Eliezer may be inexperienced in such things and, therefore, is likely to be unsure of himself, from this seemingly simple assignment he actually helps to establish another of the fundamentals and bedrock lessons of Judaism. As he is sent to find Isaac a bride, we overhear Eliezer saying:

"YHWH, God of my lord Abraham, make something happen in front of me today and show kindness to my lord Abraham" (Gen. 24:12).

What we are witnessing is the recording of the Torah's first prayer to GOD!

This first prayer is most wonderful. To whom is it directed? To YHWH, the GOD of "my lord Abraham." Who is whispering this prayer? It is Eliezer, a man of

---

1   Telushkin, 414.

no official power or rank and possibly a non-Hebrew. He is simply a respected man working for Abraham. And for what is he praying? He is praying for guidance. He asks the LORD for the judgment required to have him please his employer Abraham. As he attempts to find a suitable wife for Isaac, he asks to be given a sign that will indicate a choice that will please Isaac's father Abraham.

This first prayer is noteworthy for many reasons:

• It is spontaneous—not pre-scripted, not planned; it is unwritten and unrehearsed.

• It is unfiltered and heartfelt.

• It is a supplication for assistance to please another human being rather than being self-directed.

• Neither fame nor personal fortune nor compensation is requested or implied.

• Eliezer feels both honored to have been chosen for this task and anxious that he may disappoint Abraham.

• The prayer emanates from this man's soul and is directed toward an unknowable GOD.

• It is uttered when Eliezer deems appropriate, not at a specified time.

• It is said where Eliezer deems appropriate, not at any specified place.

Eliezer's act of praying teaches us that prayer need

not be restricted to designated times or places, nor come from prescribed texts. "Prayer is what the Talmud calls 'the labor of the heart,'" Rabbi Bradley Shavit Artson in *The Bedside Torah*. "It is answerable to the heart alone."[2]

Prayer opens the person praying to the search of a clear path that is directed toward a higher spiritual concern. It can take the form of a simple "thank you," of a plea for mental quietude or of a request for guidance, forgiveness or strength. Often, prayer is used to clear away the cobwebs that obstruct our obtaining a higher moral cognition.

This marks another nexus point where GOD and Man converge. We seem to meet the essence of GOD in the most unexpected places! Prayer is our way of opening ourselves to this essence. It is a time when we move ourselves to the side in order to "let GOD in."[3]

The Hebrew soul is never elevated when form overcomes substance or when concern for self overrides concern for others. No matter how we choose to pray, we hear this "still small voice" tell us, in response to our openness, that we should pray for aid for others, acquire guidance for ourselves and add hope, goodness, forgiveness, strength and understanding to the world around us.

Eliezer, the gentleman who whispers the first prayer, asks nothing for himself. He requests neither compensation nor promotion, but directs his prayer toward the duty that he has been asked to perform. He finds

2   Artson, 36.
3   Matt, *God and the Big Bang*.

himself by an unspecified water well, at an unspecific time of day, where he spontaneously, from his heart to GOD's presence, asks the LORD for guidance. He seeks guidance to have his actions simply be pleasing to another human being.

Now that's a prayer!

How does this compare to how we pray?

CHAPTER 23
ISAAC - THE QUIET YET STRONG PATRIARCH

The Torah's most intriguing book continues to unfold as we now meet Isaac face-to-face.

I do mean "meet" and not "listen to" as, compared with Abraham, Isaac is "cut from a different cloth." Whereas Abraham has proven to be direct and sometimes argumentative, Isaac shows us another side of humanity. Though different, Isaac also demonstrates a keen capacity to exemplify godliness.

Isaac is a quiet man, letting his calm manner and deliberate actions speak for him as he determines that a respectful demeanor based upon godly standards will characterize his chosen method for living a holy life.[1] We meet the 40-year-old Isaac in Genesis 24:63:

"And Isaac went to meditate in the field toward evening, and he raised his eyes and saw and here were camels coming. And Rebecca raised her eyes and saw Isaac, and she dismounted from the camel. And she said to the servant, 'who is that man who's walking in the field towards us'? And the servant said 'he's my lord', and she took a veil and covered herself."

Where does Rebecca, Isaac's bride-to-be, first see Isaac? Is Isaac planning the crops for next season or offering up some extravagant business potential or deciding a war strategy? No. Rebecca first sees Isaac, as does the reader, quietly meditating while standing in a field. Since the verse says that it was toward evening, Isaac could very well have been saying the mincha, or afternoon, prayers.

---

1   Telushkin, 43.

After hearing Abraham's employee, Eliezer, tell him about the "miracle" of meeting Rebecca, how does Isaac react?

"And Isaac brought her to his mother Sarah's tent. And he took Rebecca, and she became his wife, and he loved her. And Isaac was consoled after his mother" (Gen. 24:67).

That's correct—it says "he loved her"! The full interpretation here is a deep love—a love of the soul, the heart and the body. Isaac was quiet but emotionally complete.

As the episode of Isaac continues, there is a famine in the land of Canaan. Yet even with this famine, Isaac prospers *"a hundredfold in that year"* as *"YHWH blessed him"* (Gen. 26:12).

As a result of this unusual wealth, Isaac's Philistine neighbors become envious and fearful of Isaac's potential strength. In order to stop Isaac from becoming too powerful, the Philistines place mud in all of his water wells, which were dug for the family by his father Abraham. In a desert environment, clogging up wells of water is tantamount to attempted murder.

The Philistine leader Abimelek then approaches Isaac, saying: *"Go from among us, because you've become much mightier than we are"* (Gen. 26:16). How does Isaac react? Does Isaac cause a scene, draw a line in the sand, threaten, rattle his saber, declare war, cajole? No. He packs up his family and moves. In a new land, his servants dig and locate a new well with fresh water.[2] How-

---

2   Telushkin, 45.

ever, the shepherds of this new land inform Isaac that the property is theirs.

How does Isaac react to this new challenge? He packs up and moves again! He now settles his family in the town of Rehovot, where he is left in peace and continues to prosper.

Throughout these episodes, Isaac says nothing to his enemies. He never hints at a confrontation, picks up a weapon or extends a challenge. He remains focused upon what is good for his family and holy before God.

After a while, who comes to visit Isaac in his new location? None other than the king of the Philistines— yes, the same king of the very people who threatened him and made him relocate his entire family and belongings! You might say, "What gall to attempt to come talk to Isaac after what they did to him!" Or you might think that additional threats may be forthcoming.

But what does this king actually say?

"We've seen that YHWH has been with you: and we say, let there be an oath between us, between us and you and let us make a covenant with you that you won't do bad toward us as we haven't touched you, and as we've done only good toward you and send you away in peace. You are now blessed by YHWH" (Gen. 26:28-29).

Apparently, the king of the Philistines is quite impressed with the vast fortune that Isaac has amassed while undergoing these intense hardships. This king recognizes that Isaac must assuredly be blessed by his God YHWH.

How does Isaac react? Isaac chooses to accept the offer to reconcile, to choose peace over confrontation.

"And he [Isaac] made a feast for them, and they ate and drank. And they got up early in the morning and swore, each man to his brother and Isaac sent them away, and they went from him in peace" (Gen. 26:30-31).

As a cerebral man, Isaac finds ways to extend his internal peace of mind to external peace among peoples. All of those who come into contact with him sense that he is transparent in his goodness, blessed by God and, therefore, worthy of their trust.

How has Isaac acquired this silent certitude? This characteristic of quiet strength seems to have come directly from having his life threatened when he, God and Abraham together climbed the slopes of Mount Moriah, where he came to grips with the possibility of his imminent sacrifice.[3] When a young man looks death in the eye, he learns to cherish what is important in life and, most important, to choose to live his life in pursuit of peace and non-violence. The Akeda serves as a life-altering and life-defining lesson for Isaac. Life, he internalizes, is surely precious.[4]

Isaac, the second Patriarch, demonstrates that there are times when love, wisdom and a calm hand can garner success or goodwill that cannot be equaled by taking violent actions motivated by revenge or retalia-

---

3   Artson, 29-30.
4   Friedman, 120.

tion. For example, even though filling a water well with sand in a desert environment is equivalent to threatening murder, Isaac chooses to demonstrate to the rest of Mankind that sometimes confrontation is not the best choice. Isaac makes it clear that we each should choose carefully how to approach holiness. Isaac shows us a gentler side of humanity that we can take into account as we all follow our own unique, winding paths toward godliness. After everything that Isaac had been put through, "by insisting on living a meaningful life, Isaac makes it possible for each of us to do the same."[5]

In retrospect, how many times have you wished that you had chosen a different path—a path of reconciliation as opposed to one of confrontation?

We should all remember the silent strength of a loving Isaac!

---

5   Artson, 32.

CHAPTER 24

THE INCEPTION OF DECEPTION- JACOB AND ESAU

Very much like the experience of Abraham and Sarah, Isaac and Rebecca find it difficult to conceive a child. Rebecca finally becomes pregnant and God speaks directly to her, saying:

"Two nations are in your womb, and two peoples will be disbursed from your insides, and one people will be mightier than the other, and the older the younger will serve" (Gen. 25:23).

This verse establishes the basis for our next set of episodes.

The twin boys, Jacob and Esau, eventually find themselves in the unfortunate situation of having each of their parents choosing to favor a different child. Isaac favors Esau, the son of the fields who enjoys hunting, and Rebecca favors Jacob, the son of the tents who is a shepherd.

One day Esau comes in from working the field and is so hungry that he actually sells his birthright to his brother Jacob for a simple bowl of meat stew. (Some translations have this as lentils.) This transaction is truly out of balance; something does not seem right. We know from God's speech to Rebecca that the younger will rule the older, so what is the true meaning of the birthright transference? Selling a birthright to a younger brother for the price of a mere bowl of stew seems strange and has the reader wondering what this presages for the long-term relationship between these brothers. Something is wrong! What will be the real cost to Esau of giving up his birthright for such a meager price?

We do not have to wait long for an answer, as our suspicions of discord are justified and the trouble soon begins.

Rebecca overhears Isaac telling Esau to bring him some of Esau's favorite food, a special dish that Esau has prepared. After receiving the dish, Isaac would bestow his fatherly blessing upon Esau. Rebecca not only favors Jacob but, having been told directly by God that the younger will rule the older, believes that Isaac should not be blessing Esau in this manner. However, instead of sitting down with Isaac and discussing the situation, she springs into action, devising a plan to have Jacob receive Isaac's blessing instead of Esau. This plan is based upon deception, the negative effects of which she actually has no comprehension.

From this portion of Genesis forward, Man will have to contend with a continuum of negative effects that will result from this initial deception. This act of deception will be felt by many people of the bible for years and years!

Rebecca assists Jacob in deceiving his father Isaac, who is in his last years and does not see or hear well, into thinking that Jacob is really Esau. As a result of this deception, Jacob receives Isaac's blessing. Rebecca believes that this is God's true desire. The tragedy is that by choosing deception as the manner in which to proceed, she unwittingly leads everyone into a spiral of deception. She has no idea of the magnitude of the problems that she has initiated!

On the face of it, we assume that Rebecca intends no

harm as she helps Jacob to receive this blessing. We can give her the benefit of the doubt with the assumption that she may truly believe that she is acting correctly, as God indeed has directly told her that He intends the younger son to carry out his work. But she alone makes the choice to deceive Isaac, and it is a curious choice at that. Rebecca has options. She can consult with Isaac, or upon realizing the difficulty of the situation she even can "ask" God for some guidance. Yet, none of these options appears to have crossed her consciousness.

Additionally, Jacob knows the plan. He can have his mother cease with the façade and end the deception. He can suggest handling this in some other way. Yet, he does not! Therefore, Jacob also makes a choice—to participate in this ruse and aid in this charade.

The obvious point is that deception has won the day! What will be the consequences of acting in this deceptive manner in order to achieve a personally desired objective? As we will see in the next several chapters, deception often encourages more deception, the consequences of which can be severe, dramatic and tragic and may even spiral out of control.

Before we end this segment, let's also take a moment to look at how God interacts with Man during the Jacob episodes.

God has made some choices of His own. As humankind and GOD have begun to somewhat "know" each other in the progression from Adam and Eve to Noah, Abraham, the people of the Covenant and now Isaac, God seems to have decided to slightly distance Himself from Mankind for a while and let the humans try to work out their dilemmas on their own. Perhaps God is seeing whether Man can direct his maturing capabilities toward choosing to live by the standards that God has established in the Covenant and, thereby, incorporate holiness into daily life. God's conversations, interjections and interactions with Man are fewer now and, when God does directly communicate with Man, the communication takes on a different form; God is mostly suggestive or supportive, not directive!

God is letting human beings "spread their wings" and watches where the consequences of their choices might lead them. God seems to be acting as a maturing parent who, until this point, has spent quality time with his children and is now letting them experience the world for themselves. God will see whether His words have been taken seriously!

Again, developments happen quickly and God does not have to wait long for an answer. As soon as Esau finds out from Isaac that Jacob has received his blessing, Esau reacts with a new determination:

"And Esau despised Jacob, because of the blessing with which his father blessed him, and Esau said in his heart, the days of mourning for my father will be soon and then I'll kill Jacob, my brother" Gen. 27:41).

How ironic that Rebecca goes to all of this trouble to have Jacob receive Isaac's blessing in order for Jacob to become the next leader of the Hebrew people, only to have Jacob's very life threatened by Esau as an immediate and direct result of this deception! What an unforeseen, possible twist of fate it is that Rebecca's very goal in initiating her devious plan can actually backfire with Jacob's death.

Surely Rebecca does not see this coming. It seems that God's children have not learned their lessons very well. We have a terrible feeling of foreshadowing: where will the effects of this initial deception lead? Will it be controllable, or will it take on a life of its own?

We get the feeling that this whirlpool of deception[1] is about to pick up speed!

---

1   Telushkin, 52-53.

CHAPTER 25

As God communicates less directly with Man, passing the "baton of responsibility" to humankind and continuing to "watch and see" whether human beings are capable of acting in a godlike manner, Man's frailties become magnified whenever he fails to live up to God's expectations. Rebecca's choice to use deception seems to indicate that the abilities and awareness that Man acquired when eating of the Tree of Good and Bad may be insufficient for him to completely live in a godlike way.

We sense that Man grasps the general constructs of the Covenant but a more explicit road map may be necessary. Human beings seem to need help inculcating godliness and holiness into their daily actions and consciousness. We are learning just how "human" we all are. We see that we should structure our lives according to the covenantal benchmarks in order to truly have an opportunity to recapture the holiness we left behind in the Garden of Eden.

Sadly, deception is not Man's only fault that has been on display:

• Though the sons of Isaac are twins by birth, each parent chooses a favorite, the consequences of which often lead to family turmoil and sometimes have tragic conclusions.

• Adam and Eve, in eating the forbidden fruit, and Rebecca, in secretly switching her sons for the blessing, all decide to take matters into their own hands rather than to trust that God has the better plan. Their actions demonstrate arrogance, impatience and a willingness to put their personal pleasure and preferences above God's intentions.

• Human beings have not renounced violence, as witnessed by Esau's determination to kill his own brother.

Jacob's next decision adds to this list of human short-comings. Upon hearing of Esau's threat against his life, Jacob flees from his homeland and leaves his family. Even though Rebecca urges him not to leave, and even knowing that the family may temporarily self-destruct, Jacob leaves behind all that he has known. Tragically, he does so not because of a stranger's threat, but because his own brother seeks revenge.

As a result of its initiation, this deception will be tied directly to Jacob's fate and will haunt him in ways he would never have anticipated. God has given this cho-sen family a first real, unfettered opportunity to live in a godlike fashion—and what do they do with this op-portunity? They create intertwined levels of problems, insecurity, estrangements, fears and anxieties! And these develop not from an outside source or enemy, but from within their very own family unit. Here we have Man independently dipping his toes into the "waters of responsibility," and even at this early juncture he is not doing very well.

We sense that God is letting the humans stumble and struggle on their own.

Although God's interactions with human beings have become less direct, fortunately for humankind He still is "keeping in touch." We sense that God will continue to help Man apply the expanded structure of covenants and good deeds—"mitzvot"—in order for Man to increase his ability to move toward holiness. An expanded road map increasingly seems necessary for Man to understand God and to respect other people.

Jacob lives for many years fearing the revenge of his brother. Simultaneously, Esau spends these same years wishing to correct the wrong that was perpetrated upon him. Though the brothers come upon each other twice during the rest of their lives—in cordial meetings that do not result in spending any serious quality time together—imagine how it must feel for them to live in fear and estrangement from each other and from their extended family for so many years.

Just look where Man's choices are taking him as this "Plague of Deception" permeates their lives!

CHAPTER 26

JACOB MEETS GOD-YOU WILL BE SURPRISED WHERE THIS HAPPENS

Intermittent among the trials and tribulations of Jacob's life are several special, wonderful and eminently important direct encounters Jacob has with GOD. One evening shortly after Jacob leaves his family and his home in fear of his life, as he has raised the ire of his brother Esau by using deceptive means to receive their father's blessing, Jacob falls asleep and has a dream.

"And he dreamed. And here was the ladder, set up on the earth, and its top reaching to the skies. Going up and going down by it. And here was YHWH standing over him, and He said, 'I am YHWH your father Abraham's GOD and Isaac's GOD. The land on which you're lying: I'll give it to you and to your seed. And your seed will be like the dust of the earth, and you will expand to the West and the East and North and South and all the families on the earth will be blessed through you and through your seed. And here I am with you, and I'll watch over you everywhere that you'll go and I'll bring you back to this land, for I won't leave you until I've done what I've spoken to you'. And Jacob woke from this sleep and said, YHWH is actually in this place. And I didn't know. And he was afraid and he said, how awesome is this place, this is none other than GOD'S house. And this is the gate of the skies!" (Gen. 28:12-17).

Jacob not only dreams of GOD but, more accurately, "experiences" GOD. During this experience, GOD restates his blessing to Jacob—the same blessing He has given to Abraham and Isaac. Jacob is to be the next leader of his people!

When Jacob opens his eyes after that dream, he is a changed person. The significance of both what was said and how it was said to Jacob dawns upon him as he awakes. This "dream" is of utmost importance to Judaism, as Jacob's mission is reconfirmed to him directly by GOD! Jacob now knows that what Abraham and Isaac lived for was sincere and essential, and that the full magnitude and importance of this ongoing mission now rests solely with him. It is his responsibility to continue to lead the Hebrew people toward holiness.

Most important, Jacob recognizes that this responsibility has been given to him personally by GOD, and as a result he finds himself in a state of awe! Although he declares that he is "afraid," he is actually expressing that he does not know how to respond to this level of wonderment and responsibility.

Further, encountering GOD in a simple desert—a location of no particular significance—Jacob immediately realizes that GOD can be "met" anywhere and, thus, everywhere!

"He happened upon a place and stayed the night there, because the sun was setting" (Gen. 28:11).

Jacob experiences GOD in an unnamed desert and not in a designated building, in a temple, on a special mountain, near a shrine or at any other notable location. He was just in a place, in "any place"! This gives him the understanding that Man is capable of meeting GOD—or, better yet, that God can meet Man—in absolutely any place and therefore, at any time!

Jacob is awestruck to know that GOD can be and is everywhere and concomitantly may be encountered at any time. This awesomeness touches all of his senses; this experience becomes an all-encompassing "awakening" for him, both literally and figuratively. As a result, his mission becomes all-consuming.

At the same time, though, this sense of awe engenders an implied level of uncertainty and anxiety within Jacob as to how to carry out such a mission, a mission assigned by a GOD that is simultaneously everywhere and can be encountered at any time. The magnitude of such an encounter must have been daunting!

Because Jacob exemplifies Man's need for assistance in carrying out GOD's desire for us to live in holiness, it is likely that Jacob's uncertainty ultimately leads him—and us—to a spiritual yearning, a yearning that will help guide our hearts and mold our minds in the direction of living a more godlike existence.

Jacob's encounter reinforces the notion of GOD's oneness: anywhere, everywhere, at anytime!

This feels like oneness to me.

CHAPTER 27

JACOB BECOMES ISRAEL - IT IS TRULY A STRUGGLE

Jacob's next extraordinary encounter with GOD occurs when Jacob is again alone one evening. It happens after Jacob sends all of his family away from their dwellings, leaving Jacob by himself in anxious anticipation of a planned meeting with Esau.

As Jacob finds himself alone:

"A man wrestled with him until dawns' rising. And he saw that he was not able against him. And he said, let me go because the dawn has risen and he said, I will not let you go, unless you bless me, and he said to him. What is your name? And he said Jacob. And he said your name won't be Jacob anymore, but Israel, because you've struggled with GOD and with people and were able. And Jacob asked and he said, tell me your name, and he said. Why is this that you asked my name, and he blessed him there. And Jacob called the place Peni-El, because I've seen God face-to-face and my life has been delivered" (Gen. 32:25-31).

Jacob struggles with GOD and holds his own! He demonstrates to GOD that human beings are up to the challenge, and the challenge is worth the effort. Once again the manner in which Man may relate to both GOD and God has broadened and matured. Jacob's wrestling results in his name being changed to Israel.

The root of the word "Israel" is "struggle" and, in Jacob's case, it is based on this nighttime event.[1] Therefore, it is also a fitting name for the Jewish people, who have struggled throughout their history against great odds

---

1   Rendsburg, *The Book of Genesis.*

and who also have struggled to understand GOD's will and God's direction. Like Jacob, we are to hold on—to not let go but continue our struggle to understand both the relatable portions of God and the unknown essence of GOD!

Here again we see that our struggle to understand GOD is not limited to place or time. We can meet GOD whenever and wherever we are, precisely because this is where GOD is! Just as in Jacob's dream, Man's struggle is to understand a GOD that may be found in any place or at any time.

Throughout history, the Hebrew people have taken seriously their responsibility as the vassal in the covenantal structure and have unceasingly struggled to understand God's directives. Through the study of, and searching for, the holiness found in the teachings inherent in the words of the Covenant, and through striving to understand the essence and will of GOD, we hope that we can lead lives of goodness.

Each generation struggles to define morality, justice and sanctity. For itself, each generation does this within the context of its unique environment and, as a result, each generation struggles to reestablish its own commitment to "never let go" of the standards of the Torah. Struggle, in all of its forms, defines the Hebrews' underlying morality, courage and strength, which are the underpinnings of our resilient sense of hope and desire to succeed.

The question now becomes whether, like Jacob, we can find the strength to enter into this wonderful struggle of searching and studying. Do we have the fortitude to not let go and not let up? To not get frustrated, too busy, too weary? To not have our attention diverted?

Jacob's struggle teaches us to never stop searching for that which makes us holy. Only when we succumb and, therefore, diminish our struggling does hope lessen and we begin to lose touch with God. Will even Jacob, who has thus far had some experience as a deceiver — and unfortunately will soon become the deceived — have the strength to continue this struggle and succeed in breaking the chains of the Plague of Deception?

A Midrash says that "the strong man is the one that struggles with himself and wins by acting on right from wrong." So the Torah gives us the tools to help us with this struggle, which goes on within each of us.

You may have noticed that the word, "awesome" is used frequently in the Book of Genesis. It's one of my favorite words found in the Bible. "Awesome" speaks to me and connects everything, as it implies a GOD that inspires me to realize my potential. GOD's awesomeness can inspire all of us to achieve our highest expectations and acquire levels of knowledge that some people may not even know are within their grasp.

The word "awesome" implies the loftiest of goals; universal oneness; the physical, scientific and mathematical order of the universe; the splendor of the indescribably large and infinitesimally small; a GOD of consciousness, wonder and anticipation; a GOD of gravity; a GOD of Dark Matter and Dark Energy; and a God of humankind. The word speaks to the incredible improbability of life's abundance that exists as we know it.

So, like Jacob, we are awestruck, not quite able to comprehend the complexity and breadth of the situation in which we find ourselves. Jacob is not dumbstruck, but he is frozen with wonder!

"Awesome" helps me relate to a GOD that is always going to be only partly knowable to me. It is one of my nexus points in the Bible. It is a word that clearly resonates with the results of Man's attempts to know GOD's essence as well as those aspects of God to which we can relate. This word reflects the impact of "the still small voice" that we know is there but cannot completely discern. "Awesome" is awesome!

Now let's return to our struggling Jacob.

Jacob's struggle with GOD cannot be understood as a mere coincidence or contrived simply for effect. The more we struggle to study, acquire knowledge, understand and become familiar with what is truly meant by "in GOD's Image," the easier it will be for us to relate to other people and engender increasing levels of trust in God. This struggling helps to shorten the distance required for Mankind to possibly make a leap to faith in GOD. In the final analysis, complete faith is really not the issue. The main concern is our capability to achieve our highest potential for living a life of godliness.

Thus, it is up to each one of us to earn God's trust!

Simple inspiration without perspiration results in stagnation. Jacob understands the difficulty of his assignment. By saying, "How awesome is this place," he recognizes:

- the importance of his responsibility;

- the difficulty of his task;

- the simplicity of the place and time in which he has encountered GOD;

- the privilege of having God communicate directly with him;

- the "everywhereness" and "anytimeness" of GOD's presence; and

- the challenge of the assignment before him!

Jacob may be awestruck, but he is not frozen into inactivity. Jacob demonstrates his mettle through his wrestling "match" with God, which ceases as the night ends and the dawn approaches. Jacob holds his own!

So, the struggle has been enjoined; Man has seen GOD face-to-face! Jacob does not succumb in the match, but competes in a way that may earn God's respect. Although we know that Man will continue to make mistakes, God now begins to trust in Man and Man starts to have faith in GOD.

Again, being Jewish is not easy; understanding the relationship between GOD and Man is a struggle. However, it is this struggling that has provided the strength required for the Jewish people to continue to strive toward the ultimate goal of establishing righteousness and justice for all Mankind.

May our struggle only end with peace!

CHAPTER 28
DECEPTION BEGETS DECEPTION

After leaving his home in fear of his brother Esau, Jacob comes upon the city of Haran, where he meets and falls in love with Rachel. Deciding that he wants to marry Rachel, he enters into a "negotiation" with Laban, Rachel's father.

"And Jacob loved Rachel. And he said, 'I'll work for you seven years for Rachel, your younger daughter'" (Gen. 29:18).

Laban agrees, and seven years go by. On the wedding night, however, Laban switches Rachel with his older daughter, Leah, and the marriage is consummated with the wrong woman! Jacob is understandably upset and protests to Laban. Things seemingly work out when Jacob agrees to work for Laban for some additional time, and Rachel also becomes his wife.

The legacy of deception that Jacob put into motion when he chose not to stop the deception of his father Isaac is now cascading back in upon him. The Plague of Deception continues to be accompanied by terribly negative effects.

The deceiver has become the deceived!

Jacob's life has been blessed in many ways. He has communicated directly with God; he has struggled with this God, materially changing the meaning and purpose of his existence; and he has honorably worked for the bride of his choice while respectfully procuring the approval of the father of his bride-to-be. Unfortunately, however, even with all of these extraordinary experiences, he still finds himself living with the unsavory effects of the deception that he initiated.

We are struck with an almost overwhelming sense of sorrow as we watch the harsh results of these deceptions.

Jacob is in awe of God's words, and he realizes how difficult his assignment to lead the Hebrews "in GOD's Image" just might be. Yet, it seems that the deception he helped to put into motion has taken on a life of its own. He experiences its effects wherever he goes, in spite of the splendor of the other parts of his life.

Even as Jacob does eventually take Rachel as his wife, a sense of foreboding continues to surround him, manifested in the efforts to produce an heir:

- Rachel has difficulty conceiving a child.

- A despondent Leah provides the first children.

- Leah's handmaiden provides additional children.

- A lamenting Rachel allows her handmaiden to provide children as well.

- Finally, Rachel provides Jacob two children, Joseph and Benjamin.

Even as a "changed" man, Jacob learns that some of his earlier decisions and actions carry negative consequences that seriously disrupt his current existence. Jacob acted with deception, and deception is now enacted upon him. He is both the perpetrator and recipient!

What is pertinent is that when we choose unwisely, we cannot always know where our choices may lead.

CHAPTER 29
JACOB'S (ISRAEL'S) DESPAIR

We have seen Jacob become a spiritually changed man. He has wrestled with GOD and heard God's words reiterating his mission to lead the Hebrew people. He has emerged as awestruck, yet strong in character for his people. He understands his duty and the enormousness of the task.

What Jacob has not done, however, is to fully inculcate this new knowledge and direction into his own family's daily lives! While he is beginning to understand his place with God, he still does not know how to impart appropriate life requirements to the other members of his family. Even though he directly has suffered the effects of deception, he has not sufficiently taught his family how to correctly proceed with their daily lives.

Over many years Jacob creates a large family, thus, you might expect that, after experiencing one of the Bible's most varied and full lives, he would surely know how to direct his family as they mature. Instead of moving forward, however, he actually repeats some of the mistakes of the past.

For example, Jacob chooses one child to be special in his eyes over the other children:

"And Israel [Jacob] had loved Joseph the most of all his children because he was a son of his old age to him. And he made him a coat of many colors. And his brothers saw that their father loved him the most of all his brothers. And they hated him. And they were not able to speak a greeting to him" (Gen. 37:3-4).

What has Jacob done?

He has set up exactly the same set of circumstances that existed in his family when he was growing up! Isaac favored Esau and Rachel loved Jacob, and we already know the negative results that emerged from that family dynamic!

Jacob does not demonstrate his favoritism for Joseph by asking him to "prepare a special meal" as Isaac asked of Esau. However, Jacob compounds his mistake by providing Joseph with the special "coat of colors," thus ensuring that his other sons will harbor feelings of jealousy and estrangement as they are reminded of Joseph's favored status each moment of every day that Joseph wraps himself within his perceived "specialness."

As if this were not enough to initiate a potentially volatile situation, Joseph's personal behavior makes things worse. At 17, Joseph is immature; his interpersonal communication skills have not been well-honed. Joseph exacerbates the situation and displays his lack of maturity by having the audacity to analyze his brothers' dreams to his own advantage. His analyses indicate that Joseph will, in the words of the Torah, "dominate" over his brothers. "Dominate" is surely a pejorative word for Genesis to use here.[1] Joseph inflames his brothers' hatred by seeming to deliberately flaunt his favored status.

The scene is set. His brothers distrust him, dislike him and feel belittled by him.

---

[1]   Telushkin, Biblical Literacy.

One day when ten of his brothers find themselves out of their father's earshot, they discuss Joseph in increasingly ugly tones. What is actually surprising is that their discussions include the potential elimination of their brother Joseph. This favoritism by Jacob toward Joseph must have been going on for years; how else could such antipathy among brothers reach the point of considering the elimination of another brother? There is no real financial gain to be achieved, no additional title of land or treasure to be seized. Their hatred toward Joseph is simply familial jealousy that has spun out of control.

The brothers throw Joseph into a pit and, after deliberating, decide to sell him to a band of merchants. The brothers then kill a goat, take Joseph's coat of colors and stain it with the goat's blood. They intend to use the stained coat to convince their father Jacob that Joseph has been accidentally lost to them. Jacob readily believes the lie:

"We found this. Recognize it. Is that your son's coat or not? And he recognized it and said 'my son's coat.' A wild animal ate him. Joseph is torn up! And Jacob ripped his clothes and wore sackcloth on his hips and mourned over his son many days. And all his sons, and all his daughters got up to console him, and he refused to be consoled, and he said, 'because I'll go down mourning to my son at Sheol' and his father wept for him" (Gen. 37:32-35).

In one horrible moment, Jacob's life crumbles further! Jacob has just come to grips with the anxiety of fearing for his life as a result of Esau's threats, and now this news jolts Jacob to his foundation.

The Plague of Deception continues! This new horror is not just the external, hollow threat from a furious brother as the result of Jacob's initial deception of Isaac. This new deception comes full circle and is perpetrated back upon Jacob.

Jacob's new fear takes center stage.

The intensity caused by deception has risen to new heights as Jacob's sons have proceeded with the most horrible deception of all—falsely claiming the death of a child to his parent! Seemingly, the negative effects of deception can go no farther; deception can induce no greater sorrow nor produce any greater pain!

Deception has returned to Jacob's life. His choice to favor one son over the others has resulted in this ultimate deception—by his other children and directed toward whom? Ironically, it is directed back to Jacob!

Jacob has demonstrated an exceptionally high level of understanding of GOD and godliness. However, "knowing" GOD does not necessarily make for godlike behavior. It is only by transferring, translating and transplanting this understanding into wise and moral actions toward others that Man can manifest godliness and give it a demonstrable purpose. Otherwise, it is simply knowledge for knowledge's sake. We must choose to use our knowledge to move beyond mere understanding and into resolute actions.

Good thoughts and intentions are not enough. Judaism teaches the value of action and doing, not just saying and thinking! As of this juncture, Jacob has not learned this lesson well enough, and the Plague of Deception has not been stemmed by honorable actions. On the contrary, Jacob allows the exact same set of circumstances to exist in his family as those that existed within his parents' family when he was a young man.

Tragically, the lesson is not learned, and this time the consequences are much more painful for him. All of Israel empathizes with Jacob. I cannot imagine the pain of being told that your son has died!

My identification with how Jacob feels is a little too close for comfort. You may remember from the Introduction to this book that my youngest son, Noah, came very close to leaving us as a result of contracting meningitis. The agony and sheer helplessness that my wife and I felt at that time was beyond description. I truly cannot imagine how Jacob must feel when he believes that his son has actually died. The intensity of joy that my wife and I felt upon Noah's eventual recovery inversely gives me some idea of Jacob's depths of anguish.

The Torah has made its point! Everyone—Jacob, his sons and the reader—reach a point at which they collectively agree: No more of this!

The lesson is clear: identifying right from wrong is not enough. Doing what is right is what counts. Inculcating morally based actions into one's daily life is a necessity and should be so pervasive that it becomes second nature to us, as normal as breathing!

Thus, if read completely and with enough thought

to "connect the dots" between episodes, the Torah becomes a source of wonderment and guidance; the reader can be strengthened through struggle and study.

The Bible now presents a slightly different angle to the Hagar Effect. Hagar was so anxious about her son's possible dehydration that she could not see the water well that was near her all along. Now Jacob and his sons are so caught up in the effects of the deceptions that they forget that the structure of the Covenant is always there for them to use in order to begin to halt Man's moral descent.

We all must keep in mind that Judaism is not based simply upon knowing what is right or wrong. It is based upon studying life choices and then choosing to live in a godlike way. Man has it within his power to choose to perform deeds that make the world a better place from this moment forward! It is what we do today and tomorrow and each day forward that makes the difference.

Negative consequences can continue only if we allow the Hagar Effect to block our use of the "structure of optimism" available to all who choose to understand the Covenant's positive effects upon Mankind. The direction that Man takes is ultimately up to him and is determined by the choices he makes; any individual may decide, at any time of his choosing, to commit to actions that bring about a world in which tomorrow can be morally and spiritually better than today.

Judaism and its covenantal structure are therefore essentially based upon a platform of ultimate hope!

Man can choose to aid Mankind any time he wants!

Thus, we feel Jacob's despair and wish to see his spirits lifted. What will Jacob do next?

CHAPTER 30

THE ISHMAELITES - THE IRONY OF IT ALL

Studying the Book of Genesis is like peeling away the leaves of an artichoke: you enjoy each leaf, all the while approaching the essence of the artichoke's core. As I move through the episodes of Genesis, I find that there is a different lesson hidden behind each leaf and each one reveals its special, unique surprise. It has been these surprises that have been so much fun and are so illuminating.

In some chapters we encounter God directly, other times a complete episode is dedicated exclusively to Man and sometimes the impact of just one word influences the way we look at the realms of God to Man, Man to God and Man to Man! We now come upon an example of the latter, where a single word carries all of the significance. In this case it is the name of a people—Ishmaelites, the descendants of Ishmael, who was the first son of Abraham.

When Joseph's brothers place him into a pit and sell him to a passing group of merchants, the Book of Genesis is not clear as to exactly who takes Joseph out of the pit. However, there is no confusion as to which people eventually bring him to the land of Egypt. The Ishmaelites do so.

At the time that Ishmael is banished into the desert with his mother Hagar, God comes to Hagar and tells her not to worry, saying:

"I will make him into a big nation" (Gen. 17:20).

As that episode ends, for all we know the name Ishmael will never be heard of again. But lo and behold, here toward the end of Genesis we meet Ishmael's descendents, and this unexpected meeting becomes a pivotal moment in the Hebrews' entire history! When I say "pivotal," I mean just that. The truth is that if the Ishmaelites had not saved Joseph's life and brought him to Egypt, there may not be a Jewish people in existence today. Our place in history may have stopped with Jacob.

Why? Joseph would not have been in a position to provide aid to the Hebrew people during the torturous seven-year famine that was coming. Who is to say that this strange Hebrew people of the desert could have survived without the aid and guidance of Joseph?

The irony of this situation is complete when we realize that without Abraham, there would be no Ishmaelites. What is the significance of this? Ishmael is known as the father of the Arab people and a distant ancestor of Muhammad, the founder of Islam. The Ishmaelites are therefore, the direct forefathers of the followers of today's Islamic religion.[1]

That's correct! Without the Jews there would be no Islam—and without Ishmael there might not be any Jews! We may each owe our continued existence to the other! If the modern world ever needed a lesson from the Torah, this is it!

Therefore, everyone, all humankind, should recognize that we are brothers and act accordingly.

---

1   Rendsburg, *The Book of Genesis.*

Within the context of the global tensions of our world, this incredible lesson, which affects every single human on earth, emanates from a single word in this chapter: Ishmaelites. Again, this shows us the incredible power of a single word written in the Torah!

CHAPTER 31
THE NEW JOSEPH

When we first meet Joseph, he is 17 years old. Though smart, he is brash and somewhat overconfident. We are reintroduced to him 22 years later[1] as a man who has experienced a great deal in his life, has matured and exudes great wisdom.

Joseph is taken from his family, brought to a strange land and, in essence, stranded there, and forced to start a life in a country where nothing is familiar to him. Being in such circumstances would be catastrophic for almost anyone; however, Joseph proves to be "a head above the crowd." Besides finding himself stranded, he stands falsely accused of several wrongdoings that lead to his being unfairly imprisoned. While in prison, he comes upon two of the Pharaoh's former employees, who ask Joseph to interpret their dreams. Joseph complies and, subsequently, the Pharaoh himself hears of Joseph's "gift" and calls upon Joseph to do the same for him.

Joseph tells Pharaoh that his dreams predict a vast famine in the land of Egypt and warns Pharaoh that he had better begin to plan immediately for this eventuality. Pharaoh is so impressed with Joseph that he appoints him not only the national disaster manager but viceroy, second in charge of all Egypt, behind only Pharaoh himself.[2]

We now behold a Joseph who has grown from the headstrong young brother left in a pit to rise to become

1   Telushkin, 82.
2   Rendsburg, *The Book of Genesis.*

the advisor to the King of the land of Egypt! This unusual journey is explained:

"And Joseph's lord took him and put him in prison, a place where the king's prisoners were kept. And he was, there in the prison. And YHWH was with Joseph and extended kindness to him and gave him the favor in the eyes of the warden. And the warden put all the prisoners who were in the prison into Joseph's hands, and he was doing all the things that they do there. The warden did not see anything in his hand because YHWH was with him, and YHWH would make whatever he did successful" (Gen. 39:20-23).

What we learn here and throughout the chapters that follow is that YHWH is always with Joseph. Further, when Pharaoh asks Joseph to interpret his dreams, Joseph repeatedly downplays his own role. Joseph tells Pharaoh that it is God who is revealing the future and God who will supply the answers that Pharaoh needs.

What is so remarkable is that Joseph gives all of the credit for his success to his God (YHWH). Even though Joseph has triumphed over the terrible ordeals that he has encountered over many years, he recognizes that his wisdom emanates from God and has the courage to acknowledge that.

Now, when Pharaoh is ready to assign someone to the ancient version of Homeland Security Director, Food Division, he looks to Joseph:

"Pharaoh said to Joseph, since God has made you know all these things, there is no one as understanding and wise as

you. You shall be over my house, and at your mouth all my people shall conform. I shall be greater than you, only in the throne. And Pharaoh said to Joseph, see I have put you over all the land of Egypt....And Pharaoh took off his ring from his hand and put it on Joseph's hand and he had him dressed in linen garments and sent him a gold chain. And Pharaoh said to Joseph I am Pharaoh, and without you, a man will not lift a hand and a foot in all the land of Egypt" (Gen. 41:39-44).

What has happened here?

Joseph rises to a position of great power in part due to his own abilities, but also with divine support. Indeed, Joseph's relationship with God is one that has become even more sophisticated; it is now more subtle and far-reaching and less direct. It reflects God's guidance as God's hand can be "seen" in Joseph's words, actions and wisdom.[3]

In guiding Joseph, God decides that it is not necessary to directly hold Man's hand as He did in the early parts of Genesis. God "watches" His people and is "with them"; however, it is now up to Man to act upon the guidance that God provides.

Beginning with Joseph, human beings finally may be absorbing God's influence. Joseph's story provides a window for us to peer into the continuation of the subtle, but important, shift of God's less direct relationship with humanity. The wisdom and confidence with which Joseph handles himself is so apparent that even Pharaoh feels YHWH's positive effect upon Joseph and is so taken with Joseph that he assigns this "Hebrew,

---

3   Rendsburg, *The Book of Genesis.*

blessed by God" to lead the Egyptian people through this most difficult time.

Think about this for a moment.

Pharaoh, the "god" of Egypt, senses the beneficent influence of the GOD YHWH upon this single Hebrew and is so in awe of this influence that he sublimates his own position and allows Joseph to assume life-or-death power throughout his land of Egypt. Joseph's position puts him in charge of allocating the scarce food resources of the land during a famine, thus, he holds in his hands the veritable power of life and death for the nation's inhabitants.

Pharaoh must be overwhelmed by Joseph to empower him to this degree. And, again, it is equally noteworthy that Joseph, often sets aside his personal pride and gives YHWH credit for much of his success and increased responsibilities.

Through Joseph we see indications that humankind may be "getting it."

CHAPTER 32
THE REUNION IS COMING

"Because the famine was in the land of Canaan" (Gen. 42:5).

As the famine extends throughout the land, Jacob learns about a food repository in Egypt. He instructs ten of his sons to go to Egypt, procure sufficient amounts of this food and bring it back to their families in order that they might survive.

Jacob keeps home only the youngest son, Benjamin, *"in case some harm will happen to him"* (Gen. 42:4). The brothers arrive in Egypt and, with trepidation, we continue through these pages imagining Jacob's sons following this direction since, unwittingly, Jacob has corralled the necessary ingredients for a potentially tragic and volatile confrontation.

The brothers are sent to address the man who, for these past 22 years, has endured the knowledge that his own brothers cast him out from his family and, as far as they may be concerned, he is deceased. His brothers caused him to face uncalled for humiliation, possible slavery, false imprisonment and untrue accusations. Joseph has weathered these indignations with internal courage, honor, and dignity supported by the omniscient hand of God.

The irony of this upcoming meeting is that the brothers will be asking Joseph for the wherewithal necessary to save their lives, when it was their direct actions toward him that almost caused Joseph to lose his! When the now powerful Joseph eventually realizes who it is standing in front of him and for what purpose, how will he react? After all that has transpired since the time that his brothers left Joseph in the desert, what will he do now?

If you were Joseph, how would you react?

CHAPTER 33

THE BROTHERS ARRIVE - JOSEPH INITIAL TEST

"And Joseph's brothers came and bowed to him, noses to the ground. And Joseph saw his brothers and recognized them, but he made himself unrecognizable to them, and he spoke with them in hard tones. And he said to them, 'from where have you come'? And they said 'from the land of Canaan, to buy food'. And Joseph recognized his brothers, but they did not recognize him" (Gen. 42:6-8).

Joseph's initial reaction upon seeing his brothers for the first time after so many years is one of caution. He will judge for himself whether his brothers continue to live by treachery and deceit or if perhaps they have changed. Joseph decides to test their character by accusing them of being spies.

The brothers immediately answer that they are honest men and are not spies. The brothers' very next statement is, for me, one of the most poignant of the entire second half of the book of Genesis:

"And they said, 'your servants are 12 brothers. We are sons of one man in the land of Canaan, and, here the youngest is with our father today and one is no more'" (Gen. 42:13).

This is a most eye-opening moment for Joseph! After these many years, Joseph now hears directly from his brothers that not only are his father Jacob and youngest brother Benjamin still living but, equally as important, that they believe one of their brothers is "no more." Of course, that particular brother is the one standing directly in front of them!

We all can imagine the whirlwind of emotions that Joseph must be experiencing at this moment. This unexpected meeting and its unanticipated revelations probably occur quickly, with no time for Joseph to collect his thoughts. However, as has come to be typical of the Bible, there is no mention of any immediate reaction whatsoever from Joseph. As in previous episodes, the reader is expected to bring his life experience to bear in order to fully understand what he is reading. The situation truly comes alive for the reader as he interjects his reactions and feelings into this encounter. Thus, the inherent increase of tension that we assuredly feel on behalf of Joseph becomes palpable.

In order to continue to test his brothers, Joseph says:

"Do this and live, as I fear God. If you are honest, one brother from among you will be held at the place where you are under watch; and you, go bring grain for the famine in your houses. And you will bring your youngest brother to me, and your words will be confirmed, and you will not die. And they did so" (Gen. 42:18-20).

The brothers accept these conditions. As they prepare to depart, Joseph chooses Simeon as the one who will remain in Egypt until their return. He also instructs his servants to not only fill the brothers' sacks with food, but also to place the silver that they have brought with them to make the purchase of food back into their sacks.

Imagine the brothers' dread when they arrive back in Canaan, open their food sacks and find the silver inside! What must have been going through their minds, since:

- The Egyptian Lord already has accused them of being spies.

- With their possession of the silver they will assuredly also be accused of theft.

- Their brother Simeon is being held captive, with his fate unknown.

- They must somehow tell their father Jacob that not only has he lost Joseph, but the Egyptian viceroy demands that the only way for them to prove that they are honest men as well as to reclaim Simeon is to return to Egypt with Benjamin, their youngest brother.

We now arrive at one of the most memorable moments of the entire Book of Genesis!

Jacob cries out:

"I am bereaved, I am bereaved!" (Gen. 43:14).

Jacob is confronted with the culmination of all of the negative effects caused by the initial inception of the spiral of deception. He is devastated. The Hebrews and their households rely on their patriarch to supply the food they require in order to survive, but the only way to save them from starvation is by sending his beloved youngest son Benjamin to Egypt and to whatever fate awaits him there.

Jacob has reached his breaking point. His options are to let his people die from lack of food or let his soul disintegrate. The choice is Jacob's to make. It is almost as if the energies and teachings of the past several episodes of the Book of Genesis have all been leading up to this single moment.

It is at times such as these that Man is called upon to display his capacity for godlike wisdom. Jacob must be in anguish. He already has lost his favorite son Joseph. He must ask himself whether the Egyptian Lord will honor his pledge to release Simeon and, even if Simeon is released, what fate will then befall Benjamin. He may be realizing that there is a strong possibility of his losing two more beloved sons. To make things worse, upon their return to Egypt his other sons may be accused of being not only being spies but also thieves, and who knows what the consequences of that might be?

How does Jacob make a choice? Upon what basis will he make his decision? Will deception find its way into this choice?

The choice that Jacob must make is a heavy one indeed. His extended family may literally die if they do not go, and Jacob may figuratively die if they do go.

Ultimately what has Jacob (Man) learned? What will be the true results of his struggles?

CHAPTER 34
JOSEPH'S TEST - THE EFFECT UPON JACOB

What does Jacob decide? Despite his anguish, Jacob sends his sons back to Egypt!

He has chosen to risk enduring extreme levels of personal grief if anything were to happen to any of his sons and makes a decision that will be for the greater good! He places his responsibility as the leader of his people above the possibility that he may face personal hardship; he concludes that receiving sufficient food for his family is more important than protecting himself from personal pain.

The test that Joseph has established for his brothers inadvertently becomes the seminal test for Jacob. The prior actions of Jacob's sons place him in a position that requires him to make the most important decision of his life.

As Jacob makes his choice, we sense that this is a somewhat different Jacob from both the young man who helped to introduce deception into his family and the older man who has been dealing with the deceptions perpetrated upon him. This Jacob may now be closer to godliness. We do not know for sure, but can it be that the negative effects of deception are diminishing? Can it be that the Jacob who helped to initiate the Plague of Deception is morphing into the Jacob who so wonderfully struggled with God and experienced His awesomeness?

We fervently hope that the answer is "yes."

CHAPTER 35

DISCRIMINATION - WHERE YOU LEAST EXPECT IT

The tests that Joseph uses to gauge the true moral character of his brothers unexpectedly become a most burdensome test for Jacob. Wonderfully, Jacob chooses the survival of his extended family over the possibility of experiencing unimaginable personal grief.

With this inspirational choice, the testing becomes re-focused on the brothers, including Benjamin, as they all prepare to make the return trip to Egypt. They take with them double the amount of silver that they brought on their first trip in order to offer Joseph proof that they are not thieves. This sign of thoughtfulness is yet another clue that humankind is perhaps learning to make restitution for having lived in the shadows of deception.

As the brothers approach Egypt, Joseph directs one of his employees to again bring them to Joseph's home. The brothers quickly realize where they are being led.

"The men were afraid, because they were brought to Joseph's house, and they said. "We are being brought on account of the silver that came back in our bags the first time, in order to roll over us and to fall upon us and to take us as slaves" (Gen. 43:18).

We sense the brothers' increasing levels of apprehension as they arrive at Joseph's house. Their feelings of uncertainty must increase with each step.

When Joseph enters his house and confronts his brothers, four very important things occur:

1. Joseph inquires about the health of their father. The brothers respond that Jacob is alive, and it is implied that he is in good health.

2. Although Joseph is not yet certain of the true moral character of his brothers, his feelings for his family are beginning to get the best of him:

"And Joseph hurried because his feelings for his brothers were boiling, and he looked for a place to weep and came to his room and wept there." (Gen. 43:30).

This reflects the deep struggle within Joseph between two conflicting emotions: the desire to believe that his brothers are worthy individuals versus what he knows has been, at least in the past, their penchant for unsavory behavior.

3. Joseph determines that more testing is required, but only after providing his brothers with a meal.

4. Again, just when the Bible seems to be heading in one direction, the reader is moved out of his comfort zone and challenged in completely unexpected ways.

"And Joseph washed his face and went out and restrained himself and said, 'Put out bread,' and they [Joseph's servants] put it out for him and... for them by themselves and for the Egyptians who were eating with him and by themselves, because the Egyptians could not eat bread with the Hebrews, because it was an offensive thing to Egypt" (Gen. 43:31-32).

In order to feel the full impact of these verses, I had to read them several times. Here is Joseph, the second most powerful man in all of Egypt, directing his servants to set out a meal for everyone, both the Egyptians and the Hebrews. Yet the Egyptians eat their food by themselves since the partaking of a meal with the Hebrews is an "offensive" action in Egypt!

Directly in the middle of this episode where Mankind, through Joseph, seems to be unshackling from the behavioral chains of deception and engaging in ethically positive behavior, Genesis reminds us that human beings do not always *"love thy neighbor as thyself"* (Lev. 19:18) even though this "golden rule" is one of the most important principles of our ethical mitzvot. At the time, it was acceptable Egyptian policy to discriminate against the Hebrews, and in this passage it is understood that the inferior Hebrews are to be left to eat by themselves. This discrimination is so acceptable that Joseph does not protest even though he is aware that his own family is being slighted right in front of him!

Returning to the analogy that reading the Torah is similar to peeling an artichoke, here is one of those unpredictable surprises behind the leaf. As we near the end of the book of Genesis and are acquiring the feeling that perhaps, as exemplified by Jacob and Joseph, Man may be beginning to display signs of his ability to behave ethically, the text knocks us back on our heels and reminds us of Man's capacity for insensitivity toward his fellow Man.

If the Egyptians can practice discrimination against the Hebrews directly in front of Joseph, their second most powerful official, then let it be known that no one is exempt from its reach. The Torah is reminding us that discrimination can occur at any time and against anyone and that the Jewish people have experienced discrimination throughout their entire history! Therefore, the Jewish individual is to remember how harsh are the wounds of discrimination and how ugly are its scars.

The Hebrew knows this pain! Discrimination is anathema to our soul!

Thus, whenever we see an act of discrimination, it is our responsibility to remember how terribly unjust discrimination feels and how insidious are its effects — and to act to eliminate it!

Chapter 36

Joseph's Final Test

"And the men looked amazed at one another" (Gen. 43:33).

Joseph wants to believe that his brothers are honorable people, but he still is not quite convinced so he keeps them on an emotional roller coaster. As his brothers approach his home they expect to be greeted by acrimonious accusations of spying and possible thievery. Instead, to their amazement, they are treated to a meal fit for a king. They surely do not expect this!

As the brothers continue with their meal, Joseph heaps even more delight upon them:

"He commanded the one who was over his house saying 'fill the man's bags with as much food as they can carry and put each man's silver in the mouth of his bag, and put my cup, the silver cup, in the mouth of the youngest one's bag, and the silver for his grain'" (Gen. 44:1-2).

But this is the way Joseph continues to test his brothers. As they set forth on their return journey to Canaan, they must be feeling emotionally disoriented but also greatly relieved. They expected to be treated harshly and instead are provided with a wonderful meal and sufficient food to take back to their families. They do not realize that Joseph has orchestrated all of this to further test them!

For this final test, Joseph sends one of his people out to confront his departing brothers and accuse them of having the audacity to steal Joseph's personal property. Joseph's representative declares to the brothers:

"The one with whom it's found will be Joseph's servant" (Gen. 44:10).

You can just imagine the incredible, wild swings of emotion that the brothers must continually be experiencing. They are incredulous at this new request.

"Each man lowered his bag to the ground and each man opened his bag. And he searched. With the oldest, he began, and the youngest he finished. And the cup (of Joseph) was found in Benjamin's bag. And they ripped their clothes, and each man loaded his ass and they went back to the city" (Gen. 44:11-13).

Joseph's royal cup is found in Benjamin's bag. Out of shear desperation the brothers rip their clothes, unable to believe what is happening to them. These men:

- are accused of being spies;

- are forced to leave their brother Simeon behind in Egypt;

- must warily break the news regarding Simeon and Benjamin to their father Jacob;

- find that, somehow, silver had been placed in their returning food bags;

- return to Egypt believing that a robbery charge may be added to one of spying;

- are unexpectedly provided a wonderful meal and sufficient food to take back to their families; and then

- discover that Joseph's personal drinking cup has found its way into Benjamin's food bag!

At this point the brothers must be reeling! They surely are wondering, "What can happen next?" Whatever anxiety they experience on their first two meetings with Joseph must increase ten-fold as they make their way back to Joseph's house yet again! Their only hope is that perhaps Joseph will be lenient with them when they again meet face to face.

As the brothers return to Joseph's home, Judah takes the lead:

"And Judah and his brothers came to Joseph's house, and he was still there, and they fell to the ground before him" (Gen. 44:14).

"And Judah said, 'what shall we say to my Lord? What shall we speak? By what shall we justify ourselves? God has found your servant's crime... Both we and the one in whose hands the cup was found" (Gen. 44:16).

Joseph's reaction to Judah's supplication perfectly establishes the perspective for this final test of the brothers:

"And Joseph said 'far be it from me to do this; The man in whose hands the cup was found; he will be my servant and you go up in peace to your father" (Gen. 44:17).

In essence, Joseph tells them that he appreciates the honesty and candor; instead of punishing all of them he says he will simply keep Benjamin in Egypt as his servant and the rest of them are free to leave. Joseph ardently wants to believe that his brothers now possess high moral character. However, he restrains himself and waits to see how they will react to this final test.

Judah, who long ago was instrumental in leaving the young Joseph in the pit to be sold to passing merchants, now comes face-to-face with another one of the Torah's nexus moments. Will Judah (Man) and the GOD of the Hebrews find a point of consensus? Will the essence of godliness be reflected through the actions of a human being?

Judah has two choices:

1. He can take his other ten brothers home and leave Benjamin in Egypt. Since the brothers' bags are packed with sufficient supplies for their families, all that Judah will have to do upon their return to Canaan is tell his father, Jacob, that his son Benjamin had an accident or that he decided to stay behind.

2. He can confront Joseph.

If Judah confronts Joseph, will Joseph react by reopening the charges of spying and possible thievery against the brothers? Will they be arrested and placed into jail—or worse? But if Judah does not confront Joseph and returns to Canaan without Benjamin, will Jacob's heart simply break?

Judah cannot predict how Joseph will react. On what basis will Judah make this decision?

As the end of the Book of Genesis approaches, the essential question is: has humanity finally learned to embrace the spirit and structure of the Covenant? Is Man capable of holding up his side of the bargain? Is living in GOD's image just a slogan, or is it a path toward holiness? Is Man capable of repenting?

How Judah responds right then either will demonstrate Mankind's growing recognition and understanding of the significance of the many lessons in the book of Genesis, or it will reemphasize Man's frailties and shortcomings.

Will the negative effects of the Plague of Deception be passed on from one generation to the next?

Joseph awaits Judah's response! Mankind awaits Judah's decision! The whole world is watching!

CHAPTER 37
JUDAH'S RESPONSE

"Judah went over to him [Joseph] and said, 'please, my lord, let your servant speak something in my lord's ears, and let your anger not flare at your servant, because you are like Pharaoh himself'" (Gen. 44:18).

Without knowing how Joseph will react, Judah decides to confront Joseph! With the die now cast and the brothers' fate unknown, Judah explains to Joseph how he told Jacob of Joseph's requirement to have the youngest brother accompany them on their return trip to Egypt in order to clear their name, rescue Simeon and attain enough food for their families. Judah further relates to Joseph that before giving his approval, Jacob looked longingly at his son Judah and imparted to him one of the Bible's most moving sentiments:

"You know that my wife gave birth to two for me and one went away from me, and I said, he's [Joseph's] surely 'torn up' and I have not seen him since. And if you take this one from me as well, and some harm happens to him, then you will bring down my gray hair into wretchedness" (Gen. 44:27-31).

Judah tells Joseph of the anguish that Jacob, their father, still feels regarding the supposed loss of his favorite son Joseph many, many years ago. Judah concludes by saying:

"And now when I come to your servant, my father, and the boy is not with us, and he is bound to him soul to soul, it will be when he sees that the boy is not there, he will die; because your servant [Judah] offered security for the boy to

my father, saying, 'if I did not bring him to you then I will
have sinned against my father for all time. And now let your
servant stay as my Lord's servant in place of the boy, and let
the boy go up with his brothers—for how could I go up to my
father and the boy is not with me, or else I will see the wretch-
edness that will find my father" (Gen. 44:30-34).

Judah relates to Joseph that, to that very day, Jacob still has not healed the hole in his heart that was
created the moment he heard about Joseph's "death."
Therefore, since Jacob's soul is now tied to Benjamin's,
if Judah returns to Canaan without Benjamin the agony
will be overwhelming and Jacob will surely die. Coura-
geously, Judah volunteers to take Benjamin's place and
remain in Egypt, thus making it possible for his brothers to return to their families and bring home with them
the life-sustaining food.

Judah's discussion with Joseph is no confrontation
at all. It is a plea for righteousness and mercy. It is both
an acceptance of personal responsibility and a demonstration of repentance.

Judah chooses to risk his personal safety in exchange
for the security of his brothers and the future of their
families! He takes responsibility for this set of circumstances and, in so doing, places his life in jeopardy. He
clearly cannot bear to see his father's countenance upon
hearing that a second of his precious children has perished. He will be unable to endure the look of sorrow
and agony that will immediately pass over Jacob's face.
Judah apparently cannot live with the guilt of allowing
another brother to be taken from his father.

Upon receiving Judah's plea, Joseph has heard enough! The testing has ended!

"And Joseph was not able to restrain himself in front of everyone who is standing by him and he called, 'take everyone out of my presence.' And not a man stood with him when Joseph made himself known to his brothers. And he wept out loud... And Joseph said to his brothers 'I am Joseph. Is my father still alive?' And the brothers were not able to answer him, because they were terrified in front of him. And Joseph said to his brothers 'come over to me.' And they went over. And he said, 'I am Joseph, your brother, whom you sold to Egypt" (Gen. 45:1-4).

"And he fell on his brother Benjamin's neck and wept, and Benjamin wept on his neck. And he kissed all of his brothers and wept over them. And after that his brother spoke with him" (Gen. 45:14-15).

Joseph is overcome when he hears his brother Judah choose wisely and bravely. He sees that Judah is acting with wisdom and honor. Additionally, we notice that deception is nowhere to be found in Judah's answer. It has been replaced by repentance.

The joy we feel along with Joseph is all-pervasive!

We also are overcome by the joyousness of Joseph's response and the elation that assuredly the brothers must be feeling. One moment we fear that the brothers may face a terrifying future, and the next moment we are swept up in Joseph's awareness that his brothers have repented. The repentance, or teshuvah, exhibited by Judah and his brothers exemplify Man's capacity for self-examination that is needed to truly effectuate personal change and forgiveness.[1]

We share with Judah and his brothers the emotional journey from the dread of certain catastrophe to instant joy—from momentary disorientation to a sense of uplifting, followed by wonderment, finalized by overwhelming relief and elation. We have been with this extended family for a long time now, and we feel fortunate to be able to "share" this moment with them.

Joseph, too, acts honorably. He does not succumb to the temptation of exacting revenge upon his brothers. He uses neither political power nor army nor treasure against them. All who know Joseph, including the Pharaoh, understand his greatness and that the thread that connects all of the wonderful pieces of Joseph's character is his open acknowledgment of God's being with him; therefore, Joseph is great in their eyes. His sensitivity, kindness and morality spring from his awareness of God's support and ethical guidance.

Genesis concludes with Joseph, Judah and Jacob each magnificently reacting to potentially life-changing situations with wisdom motivated by righteousness and, thereby, dispelling any remnant of the Plague of Deception. Each in his own way reconnects with the teachings of the Covenant and experiences recommitting to living life according to God's standards of behavior.

The central lesson is that Man is ultimately capable of goodness—if he so chooses!

---

1   Telushkin, 87.

CHAPTER 38
GOD RETURNS

Joseph now turns to his brothers and instructs them to go back to the land of Canaan and tell Jacob, their father, the good news about Joseph. Judah asks Jacob to bring the entire family to Egypt so that they may reside there.

Jacob's response to hearing about Joseph is instantaneous:

"And Israel [Jacob] said 'So much! Joseph, my son, is still alive. Let me go and see him before I die'" (Gen. 45:28).

Now that Man has demonstrated his capacity for introspection, resulting in repentance and personal change, what happens next closes the Book of Genesis on an unmistakably optimistic tone: God decides to return to direct contact with Man!

In several of the later episodes in Genesis, God's presence is implicit, suggestive and supportive, but not direct as in the first half of the Book of Genesis. It is as if God is watching to see whether His children will take His teachings seriously.

Because of the actions of Jacob, Judah and Joseph, we know that Mankind is capable of overcoming many faults. Their godly inspired choices have led them to wonderful levels of fulfillment that were not experienced by humans earlier in the Torah.

At the end of the Book of Genesis, God reemerges in a familiar way:

"And God said to Jacob, in a nights dream, and he said 'Jacob, Jacob.' And he said 'I am here.' And he said 'I am God, your father's God. Do not be afraid of going down to Egypt, because I will make you into a great nation there. I shall go down with you to Egypt and I shall also bring you up, and Joseph will set his hands on your eyes'" (Gen. 46:2-4).

At the moment that Jacob and his people are preparing to go to Egypt, God finds it important to reinforce the Hebrew people's feelings of optimism, accomplishment and joyfulness. God is back!

He sees that Man has chosen righteousness, justice, truth and honor as the cornerstones of maturation. Man has demonstrated the ability to repent and thereby move closer to godliness. We sense that in God's return, He seems proud and comforted with Man's selection of paths leading to increased holiness. It is as if by returning to direct contact with Man, God is repeating Jacob's own words back to him: "I am here."

As the Book of Genesis concludes, I find myself sharing in the feeling of joyousness exhibited by Joseph and his brothers, in the elation experienced by Jacob and in the acceptance of repentance demonstrated by Judah. I am buoyed by the ultimate message that Man is capable of approaching godliness. The optimistic outlook as Genesis ends is mirrored at the end of the entire Hebrew Bible, or "Old Testament," which concludes with the word "good"—"Remember me for good" (Neh. 13:31)—and which serves as an "exquisite, hopeful bookend to match the opening chapter of Genesis, in which everything starts out being 'good.'"[1]

---

1   Friedman, 14.

CHAPTER 39
SO WHAT DOES IT ALL MEAN?

The Bible begins with several repetitions of the phrase, "And it was good." As our study of the Book of Genesis comes to a close, isn't it wonderful that we come full-circle with a confirmation of the potential for human goodness? We are strengthened by the hopefulness punctuated within these final episodes. We now fundamentally understand that humanity is capable of approaching godliness and that we can be instrumental in bringing holiness to Mankind.

In the Introduction to this book, I indicated that at the beginning of my journey I truly did not know what I would find at its end. I determined that I would have intelligent and honest answers for my sons' increasingly challenging questions, but I was not sure that I would like those answers. I am pleased to say that today, after my four years of study, I am more comfortable with my Jewishness than at any time in my life.

I found that modern science and religion are not incompatible, as science does not dispute faith's concept of the existence of something greater than Mankind can specifically define. Just as science does not offer the *why's* of the oneness of the physical cosmos, religion does not provide the *why's* of the spiritual No-Thingness of GOD. Thus, both are ultimately not completely knowable to Mankind. Therefore there is surely room for GOD.

I also discovered the brilliance of the Book of Genesis. This web of episodes establishes Man's ultimate goal to become "in GOD's image," yet the goal is inherently indefinable. To start man on his path towards this goal the Book then tells how Man chose to eat of the fruit of the Tree of Knowledge and acquired free will and illustrates that being free to choose our own actions can either draw us toward wanting to live in GOD's image or pull us directly away from that goal. To guide us in making righteous choices, Genesis provides the structure and road map of the Covenant and as we hold up our end of the Covenant, each of us can use our free will to struggle with life's temptations, ponder the essence of GOD and figure out how to achieve godliness. Additionally, we know that good intention must be carried through with action, as our lives are judged by the moral choices we make and that it is through our chosen actions that we ultimately demonstrate our responsibility to GOD, to other men and to ourselves. Each generation is to decide whether to recommit to the Covenant and then proceed with our search for the meaning of holiness within the context of our present-day cultural, religious and societal norms and restrictions.

I have come to respect Judaism's resiliency as it responds to this process, which results in the underlying understanding that the individual is judged upon the moral actions that he takes or will take today and in the future. Therefore, the process itself establishes the potential for tomorrow being better than today and which, thereby, places Judaism firmly upon a foundation of institutionalized hope. "To be Jewish is to hope!"[1]

I have learned that we are instrumental in this uni-

---

[1] Artson, 85.

versal give-and-take and that we should derive satisfaction from acknowledging Man's place in the universe.

Most importantly, I understand that the Torah attempts to lead Mankind toward respecting the sanctity of the individual, that holiness is delineated through taking the time to strive for justice, righteousness and dignity and that the life of each individual is precious.

Each day my studying and writing have brought me to a new level of understanding myself and my relationship with GOD. I have a deeper appreciation for the aspects of an omniscient GOD that I know I will never quite completely discern and for the aspects of God to which I am able to relate. The pervasive theological message is that the hand of GOD is always present throughout the Book of Genesis.[2]

Understanding these fundamental principles of my religion allows me to pursue increased levels of trust that can hopefully support higher levels of faith.

So let us choose to live our lives in a way that will earn GOD's faith in Man!

I wish that I had learned these things in Sunday School, Religious School or earlier in my life!

---

2   Rendsburg, *The Book of Genesis.*

APPENDIX I - THE BIG BANG

In his book, *A Short History of Everything*, Bill Bryson explains the physics of the Big Bang. Following are passages from his book that I found particularly pertinent.

"No matter how hard you try, you will never be able to grasp just how tiny, how spatially unassuming, is a proton. It is just way too small....

"A proton is an infinitesimal part of an atom, which is itself, of course, an insubstantial thing. Protons are so small that a little dib of ink like this dot · can hold something in the region of 500 billion of them, rather more than the number of seconds contained in half a billion years. So protons are exceedingly microscopic, to say the very least....

"Now imagine if you can (and of course you can't) shrinking one of these protons down to a billionth of its normal size into a space so small that it would make a proton look enormous. Now pack into that tiny, tiny space about an ounce of matter. Excellent. You are ready to start a universe....

"I'm assuming of course that you wish to build an inflationary universe. If you would prefer instead to build a more old-fashioned, standard Big Bang universe, you'll need additional materials. In fact, you will need to gather up everything there is—every last mote and particle of material between here and the edge of creation—and squeeze it into a spot so infinitesimally compact that it has no dimensions at all. It is known as a Singularity....

"In either case, get ready for a really big bang. Naturally, you will wish to retire to a safe place to observe the spectacle. Unfortunately, there is nowhere to retire to, because outside the Singularity there is no where....

"When the universe begins to expand, it won't be spreading out to fill a large emptiness. The only space that exists is the space it creates as it goes....

"It is natural, but wrong, to visualize the Singularity as a kind of pregnant dot hanging in a dark, boundless void. But there is no space, no darkness. The Singularity has no "around" around it. There is no space for it to occupy, no place for it to be. We can't even ask how long it has been there—whether it has just lately popped into being like a good idea, or whether it has been there forever, quietly awaiting the right moment. Time doesn't exist. There is no past for it to emerge from....

"And so, from nothing, our universe begins....

"In a single blinding pulse, a moment of glory much too swift and expansive for any form of words, the Singularity assumes heavenly dimensions, space beyond conception. In the first lively second (a second that many cosmologists will devote careers to shaving into ever finer wafers) is produced gravity and the other forces that govern physics. In less than a minute the universe is a million billion miles across and growing fast. There is a lot of heat now, 10 billion degrees of it, enough to begin the nuclear reactions that create the lighter elements—principally hydrogen and helium, with a dash (about one atom in 100 million) of lithium. In three minutes, 98% of all the matter there is or will ever be has been produced. We have a universe. It is a place of the most wondrous gratifying possibility, and beautiful, too."[1]

Neil deGrasse Tyson and Donald Goldsmith add to the understanding of this science in *Origins: Fourteen Billion Years of Cosmic Evolution*:

"Continuing onward with what is now laboratory confirmed physics, the universe was hot enough for photons to spontaneously convert their energy into matter—antimatter particle pairs, which immediately therefore annihilated each other, returning their energy back to photons. For reasons unknown, this symmetry between matter and antimatter had been "broken" at the previous force splitting, which led to a slight excess of matter over antimatter. The asymmetry was small but crucial for the future evolution of the universe; for every one billion antimatter particles, 1,000,000,001 particles were born...."[2]

1   Bryson, 9-10.
2   Tyson and Goldsmith, 26.

"Once scientists determined that particles and anti-particles, matter and antimatter would collide and annihilate each other....

"The combination of observation and theory suggests that an episode in the very early universe endowed the Cosmos with a remarkable asymmetry, in which particles of matter outnumbered particles of antimatter by only one part in a billion—a difference that allows us to exist today. That tiny discrepancy in population could hardly have been noticed amid the continuous creation, annihilation, and re-creation of quarks and anti-quarks, electrons and anti-electrons (better known as positrons), and neutrinos and anti-neutrinos....[3]

"Without the imbalance of a billion and one to a mere billion between matter and antimatter particles, all the mass in the universe" would have been annihilated before the universe's first seconds had passed, leaving a cosmos in which we could see photons AND NOTHING ELSE....[4]

"Still, the universe now seems disturbingly unbalanced; we expect particles and antiparticles to be created in equal numbers, yet we find the Cosmos dominated by ordinary particles, which seem to be perfectly happy without their antiparticles. Do hidden pockets of antimatter in the universe account for the imbalance? Was a law of physics violated (or was an unknown law of physics at work?) during the early universe, for ever tipping the balance in favor of matter over antimatter? We may never know the answers to these questions....[5]

"At the time when the universe was just a fraction of a second old, a ferocious trillion degrees hot, and aglow with an unimaginable brilliance, its main agenda was expansion. With every passing moment the universe got bigger as more space came into existence from nothing....As the universe expanded it grew cooler and dimmer. For hundreds of millennia, matter and energy cohabited in a kind of thick soup in which speedy electrons continually scattered photons of light to and fro....[6]

---

3   Tyson and Goldsmith, 41.
4   Tyson and Goldsmith, 42.
5   Tyson and Goldsmith, 51-52.
6   Tyson and Goldsmith, 53.

"Back then, if your mission had been to see across the universe, you couldn't have done so. Any photons entering your eye would, just nanoseconds or picoseconds earlier, have bounced off electrons right in front of your face. You would have seen only a glowing fog in all directions, and your entire surroundings—luminous, translucent, reddish white in color—would have been nearly as bright as the surface of the Sun....[7]

"As the universe expanded, the energy carried by each photon decreased. Eventually, about the time that the young universe reached its 380,000th birthday, its temperature dropped below 3000th with the result that the protons and helium nuclei could permanently capture electrons, thus bringing atoms into the universe. In previous epochs, every photon had sufficient energy to break apart a newly formed atom, but now the photon had lost this ability, thanks to the cosmic expansion (and its cooling). With fewer unattached electrons to gum up the works, the photons could finally race through space without bumping into anything. That's when the universe became transparent, the fog lifted and a cosmic background of visible light was set free....[8]

"The gestation time had ended and the Universe announced its arrival!....

"For billions of years these gases coalesced into high mass stars, each of which generates and releases sufficient energy in its core, that allows the star to support itself against gravity. Within the core of a massive star, a process of nuclear fusion occurs, which converts hydrogen into helium, helium into carbon, carbon into oxygen, oxygen into the neon and so on....[9]

"When a star can now no longer support itself against gravity by pulling a new energy releasing process out of its nuclear fusion hat, the star suddenly collapses, forcing its internal temperature to rise so rapidly that a gigantic explosion ensues as the star blows its guts to smithereens.[10]

---

7 Tyson and Goldsmith, 53-54.
8 Ibid.
9 Tyson and Goldsmith, 164.
10 Ibid.

"This creates an enormously bright and powerful Supernova explosion which effectively disperses these elements made by this process throughout the universe. This process has occurred an incalculable number of times and is responsible for producing all of the elements that we now find on Earth.

"Yes, Earth and all its life comes from Stardust."[11]

---

11   Tyson and Goldsmith, 165.

APPENDIX II - (The Covenantal Structures 2)

There are two covenantal structures in the Bible. The first, called the "Royal Grant" structure, describes God's relationship with Man through Noah and Abraham. This structure is straightforward, as God indicates to Man a rather simple understanding: "This is the moral basis upon which I would have you conduct yourself and I will guide you."[1]

It was the correct structure within which to begin the new relationship between Man and God.

However, as this relationship matured and Man's free will and ego strengthened, there existed a need for an expanded set of understandings, which resulted in the Suzerainty/Vassal (S/V) Covenant structure.

This structure, which has been examined in depth both by Old Testament scholar Walter Eichrot and author Martin Buber in his book, *I—Thou of Mankind*, reflects the maturing relationship between Man and God as it has a much more expansive set of requirements for both parties—the overlord, or Suzerain, and the under lord, or Vassal. The S/V structure established a series of requirements that had to be undertaken by both parties if the Covenant was to successfully translate into a contractual agreement between the parties.[2]

Laid out in section 4 of the Covenant, these requirements reveal the subtle genius of the Suzerainty/Vassal covenantal structure. By contractual understanding, the Torah must be relearned by each new generation and, upon hearing this reading, each generation decides whether to "re-up" and choose to commit to God's Covenant. Man is an active participant as he decides whether he wishes to agree to these edicts. It is his choice.

Since the Torah is necessarily studied by each new generation, the words of the Torah are heard within the context of the cultural, historic and social setting of the time. This is very impor-

1   Levine, *The Old Testament.*
2   Ibid.

tant, as Mankind successively hears and reads the same words that past generations heard, but with somewhat different "ears."

It is in this context that Judaism retains its vibrancy and relevance for all time! Each generation decides, within the context of that era how to best use the words of Torah in order to bring about Holiness, Justice and Righteousness within their lives and throughout the world.

An S/V Covenant is very much like a marriage:[3]

- A new Covenantal understanding is created, where no understanding preceded it.

- The relationship is entered into by choice.

- Each party agrees to support the other.

- Both parties work toward producing their best efforts.

- Each proceeds because he or she has chosen to, because he or she wishes to be a participant.

---

3   Ibid.

BIBLIOGRAPHY

Artson, Rabbi Bradley Shavit, *The Bedside Torah*. New York: McGraw Hill, 2001.

Bryson, Bill, *A Short History of Nearly Everything*. Broadway Books, 2004.

Cahill, Thomas, *The Gifts of the Jews*. New York: Nan A Talese/ Anchor Books div. of Random House, 1998.

Caldwell, Robert, cosmologist, Dartmouth College. *Scientific American*, 9/24/07.

Chaikin, Andrew, editor. *Science Today*, 11/15/2002.

Davies, Paul, About Time. New York: Simon and Schuster, 1995.

"E=MC2,"www.worsleyschool.net/science/files/emc2/emc2.html.

Fisher, Mary Pat, *Living Reflections*, 3rd ed. Upper Saddle River, NJ: Prentice Hall, 1991.

Friedman, Richard Elliott, *Commentary on the Torah*, San Francisco: Harper Collins Press, 2003.

Grist, Stan, *Fibonacci Numbers in Nature*, www.world-mysteries. com/sci_17/htm and www.stangrist.com.

Hawking, Stephen, *A Brief History of Time*, Bantam Books, 2005.

Lefkovitz, Elliot, personal notes, 2008.

Levine, Amy-Jill, *The Old Testament Vols. 1 and 2*, lecture at Vanderbilt University, DVD, Chantilly, VA: The Teaching Company Lectures, 2001.

Matt, Daniel, *God and the Big Bang*. Woodstock, VT: Jewish Lights Publishing, 1996.

McGill's Medical Guide, rev. ed. Salem Press, 1998.

McGoodwin, Michael, "Epic of Gilgamesh, Summary," prepared 2001, rev. 2006, http://mcgoodwin.net/pages/otherbooks/gilgamesh/html.

Rendsburg, Gary, *The Book of Genesis, Vols. 1 and 2*, lecture at Rutgers University, DVD. Chantilly, VA: The Teaching Company Lectures,. 2006.

Scientific American eds., *Scientific American Book of the Brain*. New York: The Lyones Press, 1999.

Telushkin, Rabbi Joseph, *Biblical Literacy*. New York: Harper Collins Publishers, 1997.

Tyson, Neil deGrasse, and Donald Goldsmith. *Origins: Fourteen Billion Years of Cosmic Evolution*. New York: W.W. Norton and Co., 2004.